STEM
LEARNING IS EVERYWHERE
Summary of a Convocation on Building Learning Systems

Steve Olson and Jay Labov, *Rapporteurs*

Planning Committee on STEM Learning Is Everywhere:
Engaging Schools and Empowering Teachers to Integrate Formal,
Informal, and Afterschool Education to
Enhance Teaching and Learning in Grades K-8

Teacher Advisory Council

Division of Behavioral and Social Sciences and Education

NATIONAL RESEARCH COUNCIL
OF THE NATIONAL ACADEMIES

THE NATIONAL ACADEMIES PRESS
Washington, D.C.
www.nap.edu

THE NATIONAL ACADEMIES PRESS   500 Fifth Street, NW   Washington, DC 20001

NOTICE: The project that is the subject of this report was approved by the Governing Board of the National Research Council, whose members are drawn from the councils of the National Academy of Sciences, the National Academy of Engineering, and the Institute of Medicine.

This project was supported by the Samueli Foundation, by Grant #1012954 from the Burroughs Wellcome Fund, and by the President's Fund of the National Research Council. Any opinions, findings, conclusions, or recommendations expressed in this publication are those of the authors and do not necessarily reflect the views of the organizations or agencies that provided support for the project.

International Standard Book Number-13:   978-0-309-30642-3
International Standard Book Number-10:   0-309-30642-6

Additional copies of this report are available for sale from the National Academies Press, 500 Fifth Street, NW, Keck 360, Washington, DC 20001; (800) 624-6242 or (202) 334-3313; http://www.nap.edu.

Suggested citation: National Research Council. (2014). *STEM Learning Is Everywhere: Summary of a Convocation on Building Learning Systems.* S. Olson and J. Labov, Rapporteurs. Planning Committee on STEM Learning Is Everywhere: Engaging Schools and Empowering Teachers to Integrate Formal, Informal, and Afterschool Education to Enhance Teaching and Learning in Grades K-8, Teacher Advisory Council, Division of Behavioral and Social Sciences and Education. Washington, DC: The National Academies Press.

THE NATIONAL ACADEMIES
Advisers to the Nation on Science, Engineering, and Medicine

The **National Academy of Sciences** is a private, nonprofit, self-perpetuating society of distinguished scholars engaged in scientific and engineering research, dedicated to the furtherance of science and technology and to their use for the general welfare. Upon the authority of the charter granted to it by the Congress in 1863, the Academy has a mandate that requires it to advise the federal government on scientific and technical matters. Dr. Ralph J. Cicerone is president of the National Academy of Sciences.

The **National Academy of Engineering** was established in 1964, under the charter of the National Academy of Sciences, as a parallel organization of outstanding engineers. It is autonomous in its administration and in the selection of its members, sharing with the National Academy of Sciences the responsibility for advising the federal government. The National Academy of Engineering also sponsors engineering programs aimed at meeting national needs, encourages education and research, and recognizes the superior achievements of engineers. Dr. C. D. Mote, Jr., is president of the National Academy of Engineering.

The **Institute of Medicine** was established in 1970 by the National Academy of Sciences to secure the services of eminent members of appropriate professions in the examination of policy matters pertaining to the health of the public. The Institute acts under the responsibility given to the National Academy of Sciences by its congressional charter to be an adviser to the federal government and, upon its own initiative, to identify issues of medical care, research, and education. Dr. Victor J. Dzau is president of the Institute of Medicine.

The **National Research Council** was organized by the National Academy of Sciences in 1916 to associate the broad community of science and technology with the Academy's purposes of furthering knowledge and advising the federal government. Functioning in accordance with general policies determined by the Academy, the Council has become the principal operating agency of both the National Academy of Sciences and the National Academy of Engineering in providing services to the government, the public, and the scientific and engineering communities. The Council is administered jointly by both Academies and the Institute of Medicine. Dr. Ralph J. Cicerone and Dr. C. D. Mote, Jr., are chair and vice chair, respectively, of the National Research Council.

www.national-academies.org

PLANNING COMMITTEE ON STEM LEARNING IS EVERYWHERE: ENGAGING SCHOOLS AND EMPOWERING TEACHERS TO INTEGRATE FORMAL, INFORMAL, AND AFTERSCHOOL EDUCATION TO ENHANCE TEACHING AND LEARNING IN GRADES K-8

JENNIFER PECK (*Cochair*), Partnership for Children and Youth, Oakland, CA
MIKE TOWN (*Cochair*), Redmond STEM School, Redmond, WA
MARGARET GASTON, Gaston Education Policy Associates, Washington, DC
LAURA HENRIQUES, Department of Science Education, California State University, Long Beach
ANITA KRISHNAMURTHI, Afterschool Alliance, Washington, DC
CLAUDIA WALKER, Murphey Traditional Academy, Greensboro, NC

JAY B. LABOV, Senior Advisor for Education and Communication, Director, National Academies Teacher Advisory Council, and Project Study Director
ELIZABETH CARVELLAS, Teacher Leader, Teacher Advisory Council

Acknowledgments

This report has been reviewed in draft form by individuals chosen for their diverse perspectives and technical expertise, in accordance with procedures approved by the National Research Council's (NRC) Report Review Committee. The purpose of this independent review is to provide candid and critical comments that will assist the institution in making its published report as sound as possible and to ensure that the report meets institutional standards for objectivity, evidence, and responsiveness to the study charge. The review comments and draft manuscript remain confidential to protect the integrity of the process. We wish to thank the following individuals for their review of this report: Barnett Berry, Center for Teaching Quality, Carrboro, North Carolina; Kathy Bihr, Tiger Woods Learning Center, Irvine, California; and Caleb Cheung, Oakland Unified School District, California.

Although the reviewers listed above have provided many constructive comments and suggestions, they did not see the final draft of the report before its release. The review of this report was overseen by Eugenie C. Scott, previous executive director, National Center for Science Education. Appointed by the NRC's Division of Behavioral and Social Sciences and Education, she was responsible for making certain that an independent examination of this report was carried out in accordance with institutional procedures and that all review comments were carefully considered. Responsibility for the final content of this report rests entirely with the authors and the institution.

We sincerely thank the following foundations in the STEM Funders

Network for their generous support of this convocation: the Burroughs Wellcome Fund; the S.D. Bechtel, Jr. Foundation; the Noyce Foundation; the Samueli Foundation; and the Charles and Lynn Schusterman Family Foundation.

We especially thank Gerald Solomon, executive director of the Samueli Foundation, for his support and encouragement throughout the planning and implementation of the convocation and to the Samueli Foundation for providing direct logistical and travel support for all participants. Michelle Freeman and Katrina Gaudier of the Samueli Foundation were most helpful in working with the committee, NRC staff, presenters, and participants in all phases of this effort. Michelle Kalista, Jan Morrison, and Meghan Sadler from Teaching Institute for Excellence in STEM also provided logistical support on behalf of the STEM Funders Network.

We also thank Monica Champaneria, Danielle Crosser, and Edward Patte, National Academy of Sciences staff members at the Beckman Center, for assisting participants during the convocation.

Contents

1

Introduction to the Convocation

Science, technology, engineering, and mathematics (STEM) permeate the modern world. The jobs people do, the foods they eat, the vehicles in which they travel, the information they receive, the medicines they take, and many other facets of modern life are constantly changing as STEM knowledge steadily accumulates. Yet STEM education in the United States, despite the importance of these subjects, is consistently falling short (see Box 1-1). Many students are not graduating from high school with the knowledge and capacities they will need to pursue STEM careers or understand STEM-related issues in the workforce or in their roles as citizens.

For decades, efforts to improve STEM education have focused largely on the formal education system. Learning standards for STEM subjects have been developed, teachers have participated in STEM-related professional development, and assessments of various kinds have sought to measure STEM learning. But students do not learn about STEM subjects just in school. Much STEM learning occurs out of school—in organized activities such as afterschool and summer programs, in institutions such as museums and zoos, from the things students watch or read on television and online, and during interactions with peers, parents, mentors, and role models. Even during their elementary school, middle school, and high school years, U.S. students spend just 18.5 percent of their waking hours over the course of each year in school (see Figure 1-1). If even a fraction of the activities and experiences that occupy much of the other 81.5 percent of their waking hours could be coordinated with the educa-

---

BOX 1-1
Identifying the Problem

During his opening remarks at the "STEM Learning Is Everywhere" convocation, Gerald Solomon, executive director of the Samueli Foundation, pointed to some well-studied problems in U.S. STEM education that the convocation was designed to address:

- Students in many other countries solidly outscore U.S. students in international comparisons of STEM learning (e.g., Kelly et al., 2013).
- Half or more of all first university degrees in Japan and China were in science and engineering, compared with just one-third in the United States. Asian and European colleges and universities are producing far more scientists and engineers than are U.S. colleges and universities (National Science Board, 2014).
- Women and minorities are underrepresented in many STEM educational programs and STEM careers (National Science Foundation, 2013).
- Growth in STEM jobs has been much faster than growth in non-STEM jobs (Langdon et al., 2011).
- STEM workers have earnings advantages at nearly every level of educational attainment (Carnevale et al., 2011).
- Solving the grand challenges that exist today in engineering, health, and other fields will require substantial contributions from STEM professionals (National Research Council, 2009, 2014a; National Academy of Engineering, 2008; Varmus et al., 2003).

Private philanthropies have an "ethical and moral imperative" to take on these problems, Solomon said. As one of the few remaining sources of risk capital available in the United States, they can support initiatives that government, businesses, and schools cannot.

---

tion they receive in school, students could emerge from their K-12 years much better prepared for the increasingly scientific and technical world in which they will live.

To explore how connections among the formal education system, afterschool programs, and the informal education sector could improve STEM learning, a committee of experts from these communities and under the auspices of the Teacher Advisory Council (TAC) of the National Research Council (NRC),[1] in association with the California Teacher Advisory Council (CalTAC),[2] organized a convocation that was held at the

---

[1]More information about the TAC is available at http://nas.edu/tac [June 2014].
[2]More information about CalTAC is available at http://www.ccst.us/ccstinfo/caltac.php [June 2014].

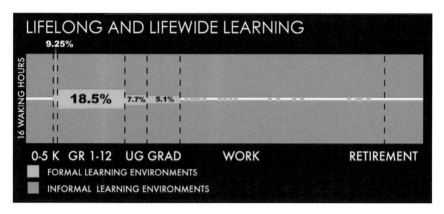

FIGURE 1-1 Estimated time spent in school and informal learning environments. NOTE: This diagram shows the relative percentage of waking hours that people across the lifespan spend in formal educational environments and other activities. The calculations were made on the best available statistics for a whole-year basis on how much time people at different points across the lifespan spend in formal instructional environments.
SOURCE: Reproduced with permission of The LIFE Center, University of Washington.

Arnold and Mabel Beckman Center in Irvine, California, on February 10-11, 2014. Titled "STEM Learning Is Everywhere: Engaging Schools and Empowering Teachers to Integrate Formal, Informal, and Afterschool Education to Enhance Teaching and Learning in Grades K-8," the convocation brought together more than 100 representatives of all three sectors, along with researchers, policy makers, advocates, and others, to explore a topic that could have far-reaching implications for how students learn about STEM subjects and how educational activities are organized and interact.

The planning committee worked from the following Statement of Task, which was approved by the NRC's Governing Board Executive Committee:

> An ad hoc steering committee will organize a convocation to explore the benefits that might accrue from engaging representatives from the formal, afterschool, and informal education sectors in California and from across the United States in strategic dialog and action planning to facilitate more deliberate connections among these three often independent communities. The emphasis of this convocation will be to foster more seamless learning of science, technology, engineering, and mathematics (STEM) for students in the elementary and middle grades that respond to new expectations and opportunities for STEM education as articulated

in the *Next Generation Science Standards* and the *Common Core State Standards for Mathematics and English Language Arts.*

Convocation participants, drawn from these three communities as well as education researchers, policy makers, professional development specialists, and funders of STEM education, will also explore how strategic connections among the three communities might catalyze new avenues of teacher preparation and professional development, integrated curriculum development, and more comprehensive assessment of knowledge, skills, and attitudes about STEM.

Based on this Statement of Task, the committee agreed that the convocation had five main goals:

1. Define the barriers to achieving more strategic, integrated approaches to STEM learning across the informal, afterschool, and formal learning platforms.[3]
2. Identify challenges and opportunities associated with developing a STEM learning "ecosystem."
3. Identify key attributes and characteristics for possible prototypes of strategic collaborations to move forward.
4. Disseminate prototypes for community uses.
5. Secure attendee commitments and devise plans of action to work on these issues for the ensuing 18 months.

The planning committee for the convocation was cochaired by Mike Town, science teacher at the Redmond STEM School in Redmond, Washington, and Jennifer Peck, executive director of the Partnership for Children and Youth in Oakland, California. The committee also included Margaret Gaston, president and executive director of Gaston Education Policy Associates (who also served as manager of CalTAC); Laura Henriques, professor of science education at California State University, Long Beach; Anita Krishnamurthi, vice president of the Afterschool Alliance in Washington, DC; and Claudia Walker, a fifth-grade teacher at Murphey Traditional Academy in Greensboro, North Carolina.

The convocation was sponsored by the Burroughs Wellcome Fund, the S.D. Bechtel, Jr. Foundation, the Noyce Foundation, the Samueli Foundation, and the Charles and Lynn Schusterman Family Foundation, which are all part of the STEM Funders Network.

---

[3]As noted by several speakers at the convocation, the terminology used in different sectors can vary. In this report, "afterschool programs" include before school, afterschool, and summer programs, which are sometimes collectively referred to as out-of-school programs.

THEMES OF THE CONVOCATION

Over the course of the convocation, the organizers of the event identified some themes that emerged from the presentations, the question and answer sessions, comments from individual participants, and the reports of breakout groups. These themes are compiled here as an introduction to the major topics that were discussed at the convocation. They should not be seen as the conclusions or recommendations of the convocation, but they represent especially promising areas for future discussion and action as identified by many participants.

Concentric circles of influences surround the individual learner, said Martin Storksdieck, then director of the Board on Science Education at the NRC (see Figure 1-2). Closest to the learner are family, friends, role

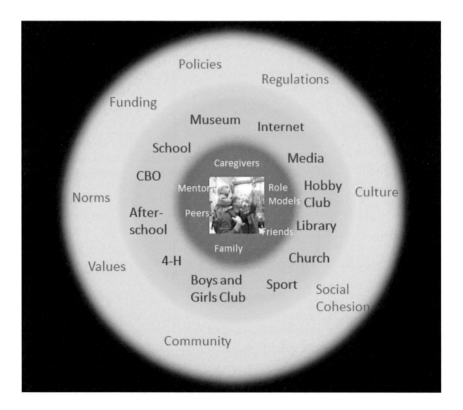

FIGURE 1-2 Individual learning is surrounded by layers of influences that have an effect on learning.
SOURCE: Reprinted with permission from Martin Storksdieck.

models, caregivers, mentors, and peers. Beyond these influences, a set of institutions seeks to teach or otherwise provide experiences to learners, including schools, museums, media, libraries, afterschool clubs, and churches. Finally, infrastructural elements such as policies, culture, communities, and values create the rules of the game for institutions as well as individuals.

The STEM learning system is remarkably diverse, Town noted, and all parts of the system have something to contribute to STEM learning. The challenge, said Henriques, is to mesh these contributions synergistically rather than duplicatively while adapting models that have worked well in one place to the culture, governance, and idiosyncrasies in other settings. In other words, they said, the goal is to create an actual integrated *system* for teaching and learning wherever learning occurs both in and outside of school (as indicated in the title of this report summary) rather than a set of uncoordinated activities.

Krishnamurthi observed that many other communities have engaged in cross-sector collaborations and that STEM learning systems could learn from those models. Solomon added that no single model of cross-sector collaboration is sufficient. Though different models may have common attributes, communities will be best served by models that reflect local cultures, environments, and stakeholders.

Peck pointed to the major changes occurring in education that provide an opportunity to align the parts of the STEM learning system. She said that the advent of the Common Core State Standards,[4] the Next Generation Science Standards,[5] new assessments that are emerging to align with these standards (e.g., the Partnership for Assessment Readiness for College and Careers [PARCC],[6] Smarter Balanced Assessment Curriculum,[7] and National Research Council, 2014b), growing interest in the social and emotional development of children, a renewed focus on effective teaching, and education finance reform are producing dramatic changes and equally dramatic opportunities. But sophisticated professional development across the entire learning system will be needed for different sectors to work together, she observed.

Different kinds of metrics will be needed for policy makers to be convinced of the value of cross-sector collaboration in producing such outcomes as persistence and having a STEM identity, said Town.

Mutual understanding, respect, and trust among the three sectors are

---

[4]See http://www.corestandards.org/ [June 2014].
[5]See http://www.nextgenscience.org/ [June 2014].
[6]See https://www.parcconline.org/ [June 2014].
[7]See http://www.smarterbalanced.org/k-12-education/common-core-state-standards-tools-resources/ [June 2014].

critical, said Walker. "We all have very similar goals," she said. "We need to find those goals and work together through them."

Finally, Peck emphasized the passion of everyone at the convocation and the surprising degree of unanimity among the participants. Even though the idea of collaboration can raise the concern that the efforts of one sector will be co-opted to meet the immediate needs of another, the commitment of the convocation participants to collaboration was inspiring, energizing, and eye-opening, she said.

ORGANIZATION OF THE REPORT

This report of the convocation summarizes the presentations, discussions, and reports to the plenary group by representatives of the breakout sessions. After this introductory chapter, Chapter 2 frames the problems the convocation was designed to address and ways of solving those problems. Chapter 3 provides four different perspectives on how better to integrate STEM learning systems. Chapter 4 examines the implications of such integration for research and for policy.

Chapters 5 and 6 summarize the main conclusions of breakout groups that discussed both particular issues and next steps in fostering collaboration among the informal, afterschool, and formal STEM education sectors. Chapter 7 compiles comments and reflections of convocation participants over the course of the event. Appendix A provides details of the convocation agenda, and Appendix B is a list of convocation attendees. Appendix C contains brief biographies of committee members and convocation presenters.

2

Envisioning the Possible

Points Emphasized by the Speaker

- Educators know the value of inquiry-based education, but without outside support, teaching tends to revert to traditional practices.
- Modern economies need the kinds of skills developed by inquiry-based education.
- The establishment of teacher advisory councils in every state and district would empower teachers to improve the education system at all levels.
- Effective partnership requires that the partners deeply respect and honor each other's unique expertise.

In his opening presentation at the convocation, Bruce Alberts,[1] former president of the National Academy of Sciences and former editor-in-chief of *Science* magazine, described three ambitious goals for science education:

1. Enable all children to acquire the problem-solving, thinking, and communication skills of scientists so that they can make wise

---

[1]The PowerPoint file for this presentation is available at http://www.samueli.org/stem conference/documents/Alberts_Moving_Forward_with_STEM_Education.pdf [June 2014].

decisions while also being productive and competitive in the new world economy. "Everybody is always trying to get your money or your vote," said Alberts. In today's complex, STEM-based world, people can make good decisions only if they know how to look for evidence and use rational thinking.

2. Generate a "scientific temper" for the United States that will ensure the rationality, openness, and tolerance essential for an effective democratic society. "We can't have a successful democracy if most people can be fooled by simple statements," said Alberts.
3. Help to generate new scientific knowledge and technology by casting the widest possible net for talent.

Educators know what science education should look like, said Alberts. Students should be exploring, hypothesizing, gathering evidence, and drawing conclusions, he said, while teachers should be acting as coaches rather than the sole authoritative source of knowledge. But it is harder to teach this way than by giving students facts to be memorized, and without outside support, teaching tends to revert to traditional approaches.

Alberts provided an example of a curriculum that enables students— kindergarteners, in this case—to engage in STEM learning:

1. Put on clean white socks and walk around the schoolyard. Because seeds stick to the fur of animals, a trait that widens dispersal, they also stick to socks, but so do many other things.
2. In class, collect all the specks stuck to socks and try to classify them. The teacher asks, "Which are seeds and which are dirt?"
3. Examine each speck with a $3 plastic "microscope."
4. Plant both the specks believed to be dirt and those believed to be seeds, thereby testing the idea that the regularly shaped ones are seeds.

To Alberts, the most important aspect of this curriculum is for teachers *not* to tell their students the answers. Teachers might direct students to draw the shapes of the objects they are investigating on paper, but they need to wait until a student suggests that the regularly shaped objects are seeds. The class then needs to agree that this is a reasonable idea and that planting the different types of objects is a good way of testing it. "This is a wonderful piece of curriculum," said Alberts, "and if it's taught well, it enables five-year-olds to think like scientists."

Many such curricula exist. "Imagine an education that includes solving hundreds of such challenges over the course of the 13 years of schooling that lead to high school graduation—challenges that increase in difficulty as the children age," Alberts stated. "Children who are prepared for life in this way would be great problem solvers in the workplace, with the abilities and the can-do attitude that are needed to be competitive in the global economy. Even more important, they will be more rational human beings—people who are able to make wise judgments for their family, their community, and their nation." The challenge, he said, is enabling teachers to be comfortable with this way of teaching and providing enough time in the school day for this kind of teaching to take place.

THE BUSINESS CASE

Business and industry would welcome this kind of education, said Alberts, because it precisely fits the workforce skills that employers say they need. These skills include

- a high capacity for abstract, conceptual thinking;
- the ability to apply that capacity for abstract thought to complex real-world problems—including problems that involve the use of scientific and technical knowledge—that are nonstandard, full of ambiguities, and have more than one right answer; and
- the capacity to function effectively in work groups and environments in which communication skills are vital.

The bad news, he commented, is that most adults have incorrectly defined what science education means for students. Adults tend to think that the object of science education is to memorize facts and be able to repeat them on tests. In his own field of cell biology, he noted that by the end of their biology classes, many high school students hate cells, because biology classes focus on the names of cell parts and their processes. For example, he referred to a seventh-grade life sciences textbook that includes the sentence, "Running through the cell is a network of flat channels called the endoplasmic reticulum. This organelle manufactures, stores, and transports materials." The self-test at the end of the chapter says, "Write a sentence that uses the term 'endoplasmic reticulum' correctly." Alberts commented, "I didn't learn about the endoplasmic reticulum until I was in graduate school, and I don't think kids need to know it. It's incredibly depressing to realize what's happening in schools."

A WEALTH OF RESOURCES

When he was editor-in-chief of *Science* magazine, Alberts prominently featured STEM education, both in his editorials and in the rest of the magazine. The magazine published two-page articles from the 24 winners of a contest for the best free science education Websites.[2] It had four special issues on education, including one in 2013 on "Grand Challenges in Science Education" (Alberts, 2013). It created a Website called "Science in the Classroom," which features scientific articles from the magazine with enough background information for students to read and understand those articles. In this way, it is helping to achieve one of the most important goals established by a framework developed by the National Research Council (2012a) that provided the vision and guidance for development of the Next Generation Science Standards: Enable all high school students to "engage in a critical reading of primary scientific literature (adapted for classroom use) or of media reports of science and discuss the validity and reliability of the data, hypotheses, and conclusions" (p. 76).

According to Alberts, one of the most urgent challenges facing STEM educators today is devising tests and other forms of assessment that will measure the knowledge and skills called for by the Next Generation Science Standards (National Research Council, 2014b; Pellegrino, 2013). It is much easier to test for science words than for science understanding and abilities. But bad tests force a trivialization of science education and drive most students, including many potential scientists, away from science, said Alberts. Good tests can be devised, but it will take many talented and knowledgeable people working together to do so. "We need to get the assessments right—quickly," he stated.

Teachers also need to be empowered, Alberts said. The U.S. automobile industry learned from the Japanese several decades ago that building a better automobile requires listening to workers on the assembly line; "ground truth is essential for wise decision making," said Alberts. But, he said, education is one of the few parts of society that has failed to act on this fact. "If we are going to keep the kind of people we need in schools and attract new people to the schools as teachers, we have to support teachers' lives in much more effective ways," Alberts said. The best science teachers need to have much more influence on the education system at all levels, said Alberts, or they will go into more lucrative and respected careers. He suggested that one valuable approach would be to have teacher advisory councils, such as the ones existing today at the national level and in California, in every state and district.

Partnerships between STEM practitioners and educators also have "an amazing power" to support teachers, Alberts observed. For example, as part of the 25-year-old Science and Health Education Partnership at the University of California, San Francisco, scientists contribute more than 10,000 hours per year, are active in 90 percent of the San Francisco

---

[2]All of these resources are available at the open-access Website http://portal.sciencein theclassroom.org [June 2014].

Unified School District schools, and benefit 21,000 students. Finally, he said, outstanding teachers have much to teach other educators about the best ways to teach.

LESSONS FOR EFFECTIVE PARTNERSHIPS

Alberts ended with three lessons he said he has learned about partnerships:

1. Any effective partnership requires that the partners deeply respect and honor each other's unique expertise. Scientists and science educators both know important things that the other group does not know. Mutual respect and understanding are the basis for working together.
2. Funding agencies must work to diminish the strong incentives for "uniqueness," which is an enemy of coherence, Alberts said. Funding agencies could help replicate proven approaches by supporting cooperation and the dissemination of good practices.
3. Partnerships flourish best when the partners can focus on accomplishing an important discrete task, which for schools, afterschool programs, and the informal sector could be to work together to create powerful approaches to STEM education. For example, he suggested, a specific activity would be to create a program that awards badges, comparable to merit badges in scouting, for STEM achievement (Alberts, 2010).

Can a child's science education impart scientific values? Alberts answered that question with a favorite quotation from the physicist Jacob Bronowski (1956):

> The society of scientists is simple because it has a directing purpose: to explore the truth. Nevertheless, it has to solve the problem of every society, which is to find a compromise between the individual and the group. It must encourage the single scientist to be independent, and the body of scientists to be tolerant. From these basic conditions, which form the prime values, there follows step by step a range of values: dissent, freedom of thought and speech, justice, honor, human dignity and self-respect.

Science has humanized our values. Men have asked for freedom, justice, and respect precisely as the scientific spirit has spread among them.

3

Achieving the Vision

Points Emphasized by the Speakers

- STEM learning ecosystems could harness the unique contributions of different settings to deliver STEM learning for all students.
- Ideas currently at the forefront of STEM education reform align particularly well with the concepts of collaboration and understanding across sectors.
- Integrated STEM education could help provide students with the skills they will need as workers and citizens.
- The opportunity currently exists to operationalize and scale up the idea of STEM learning ecosystems.
- Partnerships within schools and between schools and other organizations could help teachers take advantage of the resources available outside schools.

Four presentations at the convocation considered ways of achieving the vision laid out by Bruce Alberts in his opening remarks. Each of these presenters viewed the issue from a different perspective, reflecting the multifaceted nature of STEM learning systems, yet they converged on common approaches.

A SURVEY OF CROSS-SECTOR COLLABORATIONS

In the months before the convocation, Kathleen Traphagen, an independent writer and strategist with expertise in education and youth development, and Saskia Traill, vice president of policy and research at The After-School Corporation in New York, talked with an array of thought leaders, policy makers, funders, and practitioners to find examples of cross-sector learning collaborations. (One of the programs they studied is described in Box 3-1). Narrowing their focus to 15 initiatives

---

BOX 3-1
The Detroit Area Pre-College
Science and Engineering Program

One of the programs discussed at the convocation was the Detroit Area Pre-College Engineering Program (DAPCEP),* which annually provides more than 4,000 students in pre-kindergarten through twelfth grade with hands-on exposure to science, technology, engineering, mathematics, and medicine through in-school and out-of-school educational curricula. Operating in southeastern Michigan with its primary focus on Detroit, the nonprofit organization has a 38-year track record of nurturing and motivating historically underrepresented minorities to pursue careers in STEM fields.

By offering high-quality programming in areas including chemical and mechanical engineering, computer programming, robotics, nanotechnology, and renewable energy, DAPCEP meets a niche need in the community. "Everyone has heard about the trials and tribulation of education in the city of Detroit," said executive director Jason D. Lee, a participant at the convocation. "We are an intervention strategy in that space." For example, the organization partners with eight Michigan universities to offer Saturday science and math classes (such as "Wonders of Flight" and "Forensic Crime Stoppers") to fourth- to twelfth-graders during the school year, as well as summer camp sessions.

DAPCEP is a launch pad for many students who will be the first in their families to go to college. The program also gives teachers a chance to engage deeply with students in exciting small-group activities. "Our classrooms are those loud classrooms where students and teachers are involved in hands-on learning experimentation and opportunity," Lee said.

Detroit's regional economy needs a STEM-educated workforce to transition from its traditional automotive manufacturing roots toward a technology-based economy, and DAPCEP is helping to make that transition. Around 60 percent of students attending its summer engineering academy at the University of Michigan's Ann Arbor School of Engineering subsequently applied to and were accepted by the school. In another survey, 80 percent of DAPCEP alumni said the program prepared them for higher education and careers in STEM fields and medicine.

---

*More information is available at http://www.dapcep.org/ [June 2014].

that featured collaborations among formal K-12 education, afterschool or summer programs, and/or some type of STEM-rich organization, they studied each of the programs to derive lessons that others can use to deepen STEM learning for many more children. As they wrote in the resulting report (Traphagen and Traill, 2014), the potential is for "young people's experiences [to] connect horizontally across formal and informal settings at each age, and scaffold vertically as they build on each other to become deeper and more complex over time" (p. 6).

Concept of a Learning Ecosystem

The metaphor they used in their report—and one that was discussed extensively at the convocation (see Chapter 7)—was that of a learning ecosystem, which they defined as follows: "A STEM learning ecosystem encompasses schools, community settings such as afterschool and summer programs, science centers and museums, and informal experiences at home and in a variety of environments. . . . A learning ecosystem harnesses the unique contributions of all these different settings in symbiosis to deliver STEM learning for all children" (Traphagen and Trail, 2014, p. 4).

STEM learning ecosystems are emerging all over the country, Traphagen and Traill observed in their report. These systems are in different stages of evolution, the potential for development remains great, and people from one sector often do not know the people or organizations from other sectors. But the people with whom the researchers talked are excited to reach beyond their silos and work with others as part of a unified endeavor.

Ideas currently at the forefront of STEM education reform align particularly well with the concepts of collaboration and understanding across sectors, Traphagen noted.[1] One such idea is the emphasis on crosscutting concepts and the development of scientific practices over time. Another is the importance of interest, identity, and how these concepts can be reinforced in different settings. A third is the development of what some learning scientists call noncognitive skills, such as persistence of an academic mindset. And a fourth is the opportunity to be more intentional about providing STEM learning opportunities for girls, youth of color, and economically disadvantaged children.

Through their survey, they learned that organizations have found niches where needs exist and have expanded into those niches. For example, Traill said, an afterschool program might be able to help an education system that is struggling with science professional development for its

---

[1]The PowerPoint file for this presentation is available at http://www.samueli.org/stem conference/documents/Traphagen-Traill_Cross-Sector_Collaborations.pdf [June 2014].

elementary school teachers. Or a science museum may provide a means for keeping middle school students engaged in science.

Traphagen and Traill pointed out that all of these systems have robust infrastructures. For example, 40 of 50 states have statewide afterschool networks, and the infrastructures for many of these networks are as robust as those for schools, though they may be far less visible. Similarly, the ecosystems they studied had several common attributes:

1. They are anchored by strong leaders and a collaborative vision and practice.
2. They are attentive to the enlightened self-interest of all partners.
3. They are opportunistic and nimble.

Some of the people with whom they talked expressed concerns about cross-sector collaborations. For example, a few people were worried that partnering with formal education might mean being subsumed by it. But in general, Traphagen and Traill observed, people recognized that every organization has a mission that has to be honored, recognized, and respected as collaborations occur.

Strategies for Building Ecosystems

In their report, Traphagen and Traill looked at six different strategies to build STEM learning ecosystems.

1. Build the capacities of educators in all sectors.
2. Equip educators from different sectors with tools and structures to enable sustained planning and collaboration.
3. Link in- and out-of-school STEM learning day by day.
4. Create learning progressions for young people that connect and deepen STEM experiences over time.
5. Focus curricula and instruction on inquiry, project-based learning, and real-world connections to increase relevance for young people.
6. Engage families and communities in understanding and supporting children's STEM success.

None of the 15 ecosystems they studied had engaged in all six of these strategies, Traphagen noted in her presentation, but their efforts overlap. For example, she said, afterschool programs could provide a place for teachers to work together, try something new, and test it, which they then could take back to their classrooms. Collaborations also provided time to talk about the same thing and build trust among sectors.

Structures could be established to continue cross-sector conversations even when time is short and other tasks are demanding, Traphagen and Traill observed. For example, some ecosystems used pre-service and student teachers as educators across multiple sectors, thus fostering collaboration in an early part of an educator's career. Afterschool programs were also developing curricula and then employing teachers to work with children, which subsequently changed the way they teach in their classrooms. Out-of-school programs and STEM-rich institutions were engaging families and communities and also mounting public awareness campaigns about the importance of STEM, which is an area with great potential for further development and success. The potential forms of collaboration are extremely diverse.

Finally, Traphagen and Traill proposed a series of actions in three separate categories that they said could advance these ecosystems:

1. *Practice*

   - Get ready to scale by learning more about what works and what does not.
   - Create a community of practice for STEM learning ecosystems.
   - Examine how STEM learning ecosystems can help realize the goals of the *Common Core State Standards for Mathematics and English Language Arts* (National Governors Association Center for Best Practices and Council of Chief State School Officers, 2010), the *Next Generation Science Standards* (NGSS Lead States, 2013), and *A Framework for K-12 Science Education* (National Research Council, 2012a).

2. *Research and Evaluation*

   - Learn how to assess learning outcomes across settings.
   - Disseminate relevant research more broadly and across sectors.
   - Increase opportunities to connect research and practice across sectors.

3. *Policy*

   - Craft a policy agenda that identifies strategic levers at different levels to advance ecosystem-building efforts.
   - Take better advantage of the flexibility embedded in existing policies. For example, many funding streams offer more flexibility than is currently used in practice.

A policy agenda such as this can use strategic levers at different levels to advance the development of STEM learning ecosystems, Traphagen and Traill suggested. The result could be a shared vision, mutual understanding of the unique expertise each partner brings to the table, and better outcomes for children.

TOWARD INTEGRATED STEM EDUCATION

The Friday before the convocation, the National Academies released a pre-publication version of the report *STEM Integration in K-12 Education: Status, Prospects, and an Agenda for Research* (National Academy of Engineering and National Research Council, 2014), produced by the Committee on Integrated STEM Education. Committee chair Margaret Honey, president and chief executive officer of the New York Hall of Science, described the report's major conclusions and recommendations at the convocation.[2]

As the report describes, solving the critical problems that face societies today will require contributions from across the domains of science, engineering, technology, and mathematics, yet schools are still failing to produce the kind of learning that is applicable in the real world, Honey observed. The economy's need for routine manual, routine cognitive, and nonroutine manual skills has declined dramatically in recent decades while the need for nonroutine interactive and nonroutine analytic skills has exploded, and this trend is going to intensify in the future. Whatever jobs can be automated will be, Honey said. Driverless cars, warehouses run by machines, and even drones delivering packages are going to radically reduce the needs for humans to do these jobs, just as software has reduced the need for accountants, travel agents, and others to do routine jobs.

Traditional education is not preparing students for this future, said Honey. Young people need to learn to be creative problem solvers, to take on challenges, and to collaborate with others who have different skill sets. "Classrooms in the 21st century should look more like the environment that I run, which is a science center, than a [traditional] classroom," she said.

A Framework for STEM Integration

Instead of providing a single definition of integrated STEM learning, the committee developed a framework for STEM integration in K-12

---

[2]The PowerPoint file for this presentation is available at http://www.samueli.org/stem conference/documents/Honey_STEM%20Integration%20in%20K-12%20Education.pdf [June 2014].

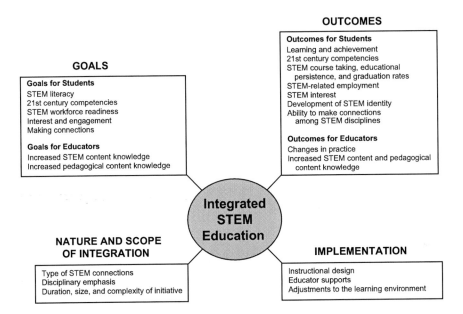

FIGURE 3-1 Integrated STEM education can be approached through a framework that includes goals, the nature and scope of integration, outcomes, and implementation.
SOURCE: National Academy of Engineering and National Research Council (2014).

education (see Figure 3-1). The framework encourages the delineation of goals, said Honey, such as what is a program trying to accomplish, what is the nature of an integrated approach, what kinds of supports need to be in place for success, how do teachers need to design their classrooms, how should they work with their colleagues both inside schools and outside of schools, and so on.[3]

Honey reported that the Committee on Integrated STEM Education developed nine recommendations, which fall into four categories and are designed to achieve the goals implicit in its framework.

---

[3]When the final version of the report was released in early March 2014, an accompanying short video that summarizes the findings of the report for general audiences also was released. That video can be viewed at https://www.youtube.com/watch?v=AlPJ48simtE [June 2014].

Research

1. Research is best when there is

    - rich description of an intervention,
    - alignment of study design and outcome measures with the goals of an intervention, and
    - the use of control groups.

2. The field—educators, program developers, researchers—could benefit greatly from a common framework for both description of an intervention and, when appropriate, for the research strategy.

Outcomes

3. Delineate the impacts on achievement, interest, identity, and persistence. Avoid the "integrated STEM is good for everything" strategy.
4. Examine the long-term impacts on interest and identify among diverse audiences.

Design and Implementation

5. Delineate a logic model, including goals, necessary supports, and outcome measures.
6. Be explicit about teaching and learning goals.
7. Use the rapidly developing cognitive and learning literature to understand learning goals and learning progressions.

Assessment

8. Rethink assessment to enable the development of high-quality assessment tools.
9. Embrace continuous improvement.

The STEM education community, both inside and outside schools, has an opportunity to build a virtuous cycle of continuous improvement, said Honey. She said a particularly promising option would be to take the framework developed by the committee and operationalize it so that different programs can be mapped onto the framework. The result could be a tool that, as Honey said, could provide much greater coherence and discipline in describing "what it is we want, how it's going to happen, what it's going to take to get there, and how we are going to know if we are successful."

THE POTENTIAL FOR AFTERSCHOOL PROGRAMS

The idea of a learning ecosystem is not new, said Anita Krishnamurthi, vice president for STEM policy at the Afterschool Alliance. Many people have been thinking about the concept for decades.[4] But, she pointed out, the opportunity now exists both to operationalize and to scale up the idea.

Nevertheless, barriers exist and need to be overcome, according to Krishnamurthi. The different settings in which science education can occur have different cultures that cannot be ignored. Instead, the differences need to be turned into an advantage, said Krishnamurthi. Each setting has a unique potential to contribute to the larger ecosystem (Afterschool Alliance, 2011b). "It's happening in small packages," she said. "The challenge is how to grow this."

Krishnamurthi said each sector lacks knowledge about the others. For example, the afterschool sector has an image problem that it is trying to rectify. Large afterschool providers such as 4-H and the YMCA have adopted STEM programming as a flagship effort within their programs, and citywide and statewide afterschool networks have come into existence to support educational objectives. People in the formal education sector now can be confident that afterschool providers can deliver on their promise, said Krishnamurthi, noting "the afterschool sector is becoming very sophisticated, savvy, and capable."

In a recent survey done by the Afterschool Alliance, nearly all of more than 1,000 afterschool program directors and staff said that it was important for afterschool programs to offer STEM programming as part of a larger comprehensive effort (Afterschool Alliance, 2011a). Yet a recent Nielsen survey found that only 20 percent of households have children enrolled in afterschool STEM programs (Change the Equation, 2013). The untapped potential to take advantage of what the afterschool community can offer is huge, said Krishnamurthi.

At the same time, evaluation research has been proceeding rapidly. A study on defining youth outcomes for afterschool STEM learning (Afterschool Alliance, 2013) found that these programs could deliver many of the goals and outcomes identified in the integrated STEM report (National Academy of Engineering and National Research Council, 2014). "Afterschool programs are a place where this kind of integrated learning and teaching can occur—is already occurring. [These] programs and providers are ready to provide this kind of learning, and we should help them do more and do better," she said.

---

[4]The Powerpoint file for this presentation is available at http://www.samueli.org/stem conference/documents/Krishnamurthi-Walker_Integrating_Afterschool_Platforms.pdf [June 2014].

The Need for New Policies

Policies will need to change to enable greater integration, Krishnamurthi said. Today, most of the resources and expectations for student outcomes are organized around what schools can deliver; which she said has a variety of consequences. For example, science centers often do professional development for teachers and sometimes for afterschool educators, but rarely do they provide professional development to both groups at the same time, usually because afterschool programs do not have the money that schools have to pay for such activities.

Another example she presented involves student data. Privacy laws are important to protect students, but they also can inhibit sharing of information between school teachers and afterschool providers. "If Mary is really struggling with wave theory in physics . . . and the afterschool provider knows that she loves music and is playing with beats and harmonies and resonance, wouldn't it be wonderful if we could make those connections in a very explicit way?" asked Krishnamurthi. "That's what this kind of thinking can engender, but policies need to be changed for us to be able to do that."

A shared vision for the entire STEM learning system would distribute responsibility for learning among all of the sectors, she noted. This responsibility could encompass not only academic achievement, but also the development of identity, interest, curiosity, and passion.

According to Krishnamurthi, leadership will be essential to make this happen, but one single charismatic leader is not enough. Rather, many players in the system need to be involved in crafting initiatives and policies, said Krishnamurthi, as has been happening in cities such as Providence and Nashville, and larger state efforts such as one in Ohio. People are asking how to make collaboration a central goal and how to incentivize people to come together and share resources in the best interests of students. She gave as an example that the Afterschool Alliance has been working with the Association of Science-Technology Centers to bring science centers and afterschool providers closer together to work on professional development. In addition, the National Girls Collaborative Project has shown that even small mini-grants can bring people together.

"We can't wait for the policies to be put into place before we start working," said Krishnamurthi. "If this is to become systemic and sustainable in the long term, we all have to advocate for this kind of thinking, find champions at the city, state, and federal levels, and ultimately change policies so that this becomes a way of life and not the exception that is driven by one charismatic leader here and there."

DISCUSSION ABOUT OPPORTUNITIES FOR
STEM IN AFTERSCHOOL SETTINGS

In the discussion session following Kristhnamurthi's presentation, Ryan Collay, director of the Science and Math Investigative Learning Experiences Program at Oregon State University, noted that the "elephant in the room" is assessment. He said academic achievement is often considered the sole measure of success, but afterschool programming is focused on a variety of outcomes, not just academic achievement. In particular, afterschool programs seek to build identity, engagement, persistence, and other attributes. "I'm worried about afterschool morphing to become more school-like," he said.

Krishnamurthi responded that more conversations about these attributes are occurring. Even among policy makers and corporate leaders, recognition is growing about the importance of outcomes other than academic achievement, she observed, and the challenge is to measure these outcomes of afterschool experiences. Afterschool programs need a way of showing that they are delivering what they are promising, she said, and such assessments need to become more robust and widespread. "It's a long, slow slog—we're not going to get there overnight. But I think the movement has begun," she said. In her view, corporate leaders in particular need to be more vocal about workforce needs for the movement to gain traction.

Krishnamurthi also pointed to research that has helped reveal the importance of STEM identity. For example, Tai et al. (2006) demonstrated the importance of having an interest in STEM subjects by the eighth grade. Eighth graders with average grades who were interested in STEM were more likely to go into a STEM field than eighth graders with good grades who did not have an interest in STEM subjects. "We just need a few more of those kinds of powerful studies to make the case to policy makers," she suggested.

In addition, Traill called for generating more standardized survey information across sectors and for greater use of the survey information that already exists. For example, the National Assessment of Educational Progress collects a lot of information about STEM activities outside of school, and the field has not been making enough use of those data. Martin Storksdieck added that many research projects are currently under way to address the issue and that a solid body of research is being compiled.

A PERSPECTIVE FROM SCHOOLS

STEM teaching in many elementary schools remains very traditional, noted Claudia Walker, a fifth-grade mathematics and science teacher at Murphey Traditional Academy in Greensboro, North Carolina, because teachers tend to teach the way they were taught. It is not that they do not want to teach in a more engaging way, Walker said; rather, they do not know how to teach that way.

The families served by many schools do not have the resources to take their children to museums or give them experiences that will spark their interest in STEM subjects. Again, it is not that they do not want to. As she noted, "Parents are always asking me, 'Mrs. Walker, what programs do we have? What can we do? What summer programs are there?'" But many summer programs are for middle and high school students, not for elementary school students, Walker pointed out, even though younger students need good science teaching, too.

Elementary school teachers face an especially great challenge when teaching students from disadvantaged backgrounds, Walker said. They have to both build background knowledge and make students realize that science is fun and not just a list of words to memorize.

BOX 3-2
OC STEM

The Orange County STEM Initiative (OC STEM)* in Southern California is a thriving example of a local partnership integrating three circles of STEM learning and teaching efforts—formal K-12 education, afterschool programs, and science institutions. OC STEM, which is also a regional network in the statewide California STEM Learning Network, has three goals: (1) to equip all students in the county with the science, mathematics, and critical-thinking skills to become competitive leaders in STEM fields; (2) to give educators the tools and support to teach those students well; and (3) to help build in Orange County the most competitive STEM workforce not just in California but in the entire United States. Although the county is home to many innovative high-technology and biomedical firms, it is not producing enough STEM-proficient students to maintain its competitive edge in the future.

The OC STEM collaboration encompasses students, parents, teachers, businesses, and funders ranging from the Samueli Foundation to Boeing. Its key implementation partners include the Discovery Science Center, an informal science institution in Santa Ana that is the largest nonprofit educational resource in the county; the Orange County Department of Education; THINK Together, a nonprofit that provides afterschool programming; and the Tiger Woods Foundation.

OC STEM kicked off in 2012 and now operates at more than 200 sites, said Gerald Solomon, executive director of the Samueli Foundation, which staffs the initiative. The program reaches more than 10,000 students each year with a variety of in-school, out-of-school, and virtual learning activities. Orange County has roughly 500,000 public K-12 students.

The Power of Partnerships

Partnerships, both within schools and between schools and other institutions, can play a big role in addressing these challenges, Walker stated. (Box 3-2 provides an example of an especially active partnership that was featured at the convocation.) She said elementary schools need people who will challenge their colleagues to try new things and think in new ways. Elementary school teachers also need colleagues interested and skilled in STEM teaching who will stay at that level and not go to a middle school or high school.

Elementary schools tend to think that afterschool and summer programs are separate entities, but it does not have to be that way, Walker said. She described her work with faculty members and students at a nearby university, including pre-service teachers, to provide the students in her school with experiences that they never would have had otherwise. Walker applied for grants that would help her bring materials and profes-

Partner members complement each other in bringing their own resources and know-how to the collaboration. For example, the Discovery Science Center houses more than 120 stimulating interactive exhibits within its 59,000 square foot museum site, but it has also long been committed to enriching science education in the community through outreach and field trip programs. As part of OC STEM, for schools lacking teachers with strong STEM expertise, the Discovery Science Center offers a series of hands-on, inquiry-based teaching activities using do-it-yourself science kits based on its Future Scientists and Engineers of America project (www.fsea.org). The museum provides materials, curricula, and professional training to front-line teaching staff who want to implement these programs, said Janet Yamaguchi, the center's vice president of education. Discovery Science Center staff also have made "pop in" classroom visits to observe and coach teachers not just on how to use the activity kits with competence and confidence but also with the enthusiasm that gets students fired up. "We were excited to see that that system worked," Yamaguchi said.

This year, OC STEM is integrating more business participation into its model through its STEM Connector—kind of an "eHarmony of education," Solomon said—that connects STEM professionals who volunteer time with in-school and afterschool activities and events. The goal for 2014 is to make 10,000 connections by December.

---

*More information is available at http://ocstem.org/ [June 2014]. The PowerPoint file for this presentation is available at http://prezi.com/e3lkkkakc3a0/?utm_campaign=share&utm_medium=copy&rc=ex0share [June 2014]. A video describing OC STEM that also was presented during this session is available at http://youtu.be/HOAn7xw__ko [June 2014].

sional development into her school, which also enabled her to take teachers out of the school to conferences and STEM-related events. She also has been working with principals and superintendents to make sure that they are supporting new practices. Science is tested in the fifth grade in North Carolina, which means that it receives more attention from administrators and teachers. Walker observed that this can result in principals being focused on end-of-year test results, but for those who are more flexible, it also provides opportunities to support learning at a high level. For example, noted Walker, science can be taught across the curriculum and can support the rest of the curriculum.

Walker said teachers need to be empowered to give their students socks and let them walk in the playground, referring to Alberts' example (see Chapter 2), or to walk around a city and think about the ecosystem in which they are living. The informal sectors can help provide students with those experiences, because they do not have many of the constraints of formal education, she stated. When students are not successful, the system has failed, not the student, said Walker, who added, "I challenge all of you to think about that—think about our students and our ultimate goals."

DISCUSSION ABOUT THE REALITIES OF
STEM EDUCATION IN FORMAL SETTINGS

During the session that ensued after comments from the discussants, Kenneth Hill, president and chief executive officer of the Chicago Pre-College Science and Engineering Program, Inc., pointed to the amount of time it takes teachers to change. "You can't do it in a week," he said. In a program developed with the Chicago Museum of Science and Industry, the assumption was that teachers need 90 hours of professional development spread over multiple years.

Hill also noted that when parents are involved with their children in STEM activities, children get a message about the importance of STEM subjects. "When children were doing science activities with their parents on Saturdays, they began to recognize that their parents value STEM, which then translated into improved student achievement Monday to Friday," he said.

4

Implications for Research and Policy

Points Emphasized by the Speakers

- Greater understanding of the learning process can inform the design of cross-setting learning.
- Diversity is a hallmark of a robust learning ecosystem.
- The leaders of afterschool and informal learning institutions have already done extensive work on the issues associated with cross-sector collaboration.
- Leadership, incentives for collaboration, and diverse advocates are important tools for policy and advocacy in cross-sector collaborations.

As noted in the presentations summarized in the previous chapter, realizing the vision at the heart of the convocation has implications for both research and policy. Two speakers explored those implications in depth, while a series of breakout groups (see Chapters 5 and 6) examined them more broadly.

IMPLICATIONS FOR RESEARCH

Over the past several decades, the learning sciences—which encompass the cognitive sciences, the developmental sciences, artificial intelligence, and the brain sciences and neurosciences—have been developing detailed accounts of learning to understand how, when, and why it happens, noted Bronwyn Bevan, director of the Research and Learning Institute at the Exploratorium.[1] One key understanding that has emerged is that learning is deeply contextual—it is a process that takes place across specific times and settings (Bransford et al., 2006). Another important conclusion, she said, is that learning is "life wide, life long, and life deep" (Banks et al., 2007). Values, beliefs, interests, and understandings are a resource for learning. They profoundly shape how one approaches and engages learning opportunities. These values, beliefs, interests, and understandings are developed in many places—in families, in communities, in all the interactions a person has with others. Bevan said they also fluctuate over time and may evolve into sustained "lines of practice" (Azevedo, 2011).

Bevan has been doing a literature review of the research on cross-setting learning with two colleagues, and she reported on several design principles they have identified (Penuel et al., 2014):

1. Draw on values and practices from multiple settings.
    a. Identify and integrate diverse values of all stakeholders.
    b. Identify practices in one setting that can be used as a resource to support learning in another setting.
2. Structure partnerships to encompass the goals of all stakeholders.
3. Engage participants in building stories, imaginative worlds, and artifacts that span contexts and that facilitate meaning-making across contexts.
4. Help youth identify with the STEM learning enterprise.
    a. Provide opportunities to contribute to authentic endeavors.
    b. Name youth as contributors or potential contributors.
6. Use intentional brokering to facilitate movement across settings.
    a. Prepare educators to play roles as brokers.
    b. Prepare parents to play roles as brokers.

According to Bevan, a prominent issue that arises in applying these principles is whether providing access is sufficient for equity. Research indicates that equity requires ongoing, multiple opportunities to do and learn STEM, she said, with opportunities for redundancies and variation. It also suggests that STEM education should be introduced as the best means for solving problems, challenges, and questions that have meaning to the learner. Finally, learning activities need to leverage children's familiar personal, family, and cultural resources and routines, according

---

[1]The PowerPoint file for this presentation is available at http://www.samueli.org/stem conference/documents/Bevan_Cross-Setting_Learning.pdf [June 2014].

to Bevan. "By making STEM education activities or programs familiar to kids, you are inviting them to come in from a position of strength," she said.

As one example of this final point, Bevan mentioned the work of Emdin (2011) with African American students in the Bronx on rap ciphers as discourse patterns. Educators familiar with rap ciphers, including overlapping speech, heightened emotions, and gestures, can learn to identify and leverage, rather than shut down, student engagement in STEM discussions in classrooms. She also mentioned work at the Exploratorium to take advantage of play in afterschool programs for STEM learning. "It's really important to leverage play," she said. "Play is a developmental resource. All humans and animals learn through play."

She noted that Penuel et al. (2014) also identified seven infrastructural elements for cross-setting learning:

1. Establish programs and individuals who can broker and support students in key transition moments, such as the transition from middle school to high school or from school to afterschool.
2. Create strategies and systems that can recognize and make visible young people's accomplishments from one setting to another. (Badges for achievement are one such system.)
3. Establish programs that connect youth with professionals and workplace or public settings.
4. Establish programs that provide classroom educators with opportunities to work with students in different low-stakes settings and contexts.
5. Use social media to link people and practices across disparate settings.
6. Intentionally relate learning opportunities in formal, informal, and afterschool settings in ways that make apparent to all stakeholders how all learning opportunities reinforce and expand young people's interest, understanding, and commitment to STEM subjects.
7. Create professional development that works across institutional boundaries to engage educators with how their efforts can collectively support interest, capacities, and commitment.

Bevan also issued several cautions. Learning ecosystems already exist, she said. They are populated by people and by institutions and are not simple. They have evolved over time as a product of complex interacting systems. The task then becomes to optimize, improve, and coordinate ecologies. She also noted that diversity is a hallmark of robust ecosystems. For this reason, diversity is an asset rather than a liability in

STEM learning systems. "A big question for us is how we are going to expand diversity of learning opportunities," she stated.

STEM learning systems need to avoid a cultural deficit model, she said, in which the norms of one culture are imposed on all students. "What is STEM, who is STEM, how do we talk about STEM, where does STEM happen?" asked Bevan. These are all critical questions in thinking about opportunities to build STEM interest.

Finally, she suggested the idea of cultivation as a metaphor. By starting with children's interests, peer groups, and strengths, STEM learning opportunities can deepen and extend their experiences.

IMPLICATIONS FOR POLICY

Educators at all levels have a huge amount of work to do to understand and figure out how to implement all the changes needed for effective cross-sector collaboration, said Jennifer Peck, executive director of the Partnership for Youth and Children. Fortunately, she noted, afterschool and expanded learning providers have been focusing on the issues associated with these changes.

Peck referred to the "four Cs" of policy work: collaboration, critical thinking, communication, and creativity. All four of these Cs can be seen in the activities afterschool programs do with students, Peck noted. Students are doing research in their communities, working in teams to find things out, and making presentations where they use technology and multimedia tools. Because of the sophistication of some of this afterschool programming, the education community is more open about including out-of-school providers in the conversation. "The thing I ask myself is how are we maximizing this opportunity and not doing it just in piecemeal ways," Peck said.

Leadership is one important element in policy and advocacy for cross-sector collaborations, said Peck, but leaders who are willing and eager to collaborate are rare. "It is a very natural tendency for systems and leaders of systems to work in their own boxes," she said. "Yes, there are policies and rules and regulations that can perpetuate this, but in a lot of cases it is really organizational culture and habits . . . that keep us in silos."

A second important policy tool that Peck pointed to is to incentivize collaboration at all levels. Collaboration is hard and can require doing things differently, she said. It can require sharing of resources and sharing of credit. It also can require policy incentives to collaborate more effectively. As an example, she noted that the 21st Century Community Learning Centers Program provided a funding stream for afterschool programs, and some states, like California, gave priority to grants that featured cooperation between schools and community organizations.[2]

But, she warned, policy opportunities also can be forgone. For example, a major missed opportunity to promote collaboration between school systems and partners occurred when the School Improvement Grants, which provided a large infusion of support for low-performing schools, provided little policy guidance about how to turn around performance.[3] As a result, schools tended to do the easiest things rather than take actions with a basis in research or best practices, said Peck. In California, for example, which has 4,500 publicly funded afterschool programs, many of which are located at the same schools that had grants, the funds generally were not used to forge connections between afterschool programs and the school day.

A third tool Peck identified is the collective role of diverse advocates. For example, when leaders and systems are doing good things, advocates should recognize their work and give them credit for having the courage to deepen and sustain partnerships. "Public recognition is an incredibly powerful motivating factor for policy leaders at a variety of levels," said Peck. "It is important to weave this into any strategy around policy change."

Multiple sectors and partners play a role in making sure that students can be successful. "Everybody has to wear an advocacy hat," said Peck. "We have to embed this concept into policies and guidance to move us in this direction, because we can't just assume it's going to happen." In California, for example, the Partnership for Youth and Children is working with the state department of education to develop and implement in state policy a definition of high-quality expanded learning opportunities. Such a definition could apply not just to afterschool programs, but also to the wide range of resources that can be used for out-of-school programming, Peck said.

The partnership is also collaborating with the state education agency on developing guidance for districts and for out-of-school time providers around how to communicate and collaborate around the new education standards. "There's a lot of confusion, at the local level, about how to do that well," Peck observed.

Finally, work is under way to develop concrete tools and information that people at the local level need to understand what the implementation of new standards means for out-of-school partners. Peck said,

[2]More information about the program is available at http://www2.ed.gov/programs/21stcclc/index.html [June 2014].

[3]More information about the grants is available at http://www2.ed.gov/programs/sif/index.html [June 2014].

"This is true for the Common Core, and it's going to be just as true for
the Next Generation Science Standards, where we have an even bigger
opportunity." Discussions about policy often get stuck on the barriers to
collaboration, said Peck. More time needs to be spent determining what is
a real barrier and what is a perceived barrier and how creative solutions
can overcome both, she said, noting that "strong leadership is absolutely
essential to all of this."

DISCUSSION

During the discussion session, Justin Duffy, a STEM specialist with
World Learning in Brattleboro, Vermont, asked how the lessons derived
from the convocation could be applied across socioeconomic and racial
lines so that all schools and students benefit from the integration of STEM
learning. Bevan emphasized the importance of examples that can serve as
visual talking points and the need to have conversations across sectors.
"We don't have those at this point. We have pockets of activity, but it is
not integrated into the mainstream conversation about STEM education,"
she responded.

Christopher Roe, chief executive officer with the California STEM
Learning Network in San Francisco, asked about the role of "backbone"
organizations in supporting and sustaining an integrated approach. Peck
referred to the role of intermediaries at the local or regional level in facilitating and brokering conversations and in directing attention to the policy
arena. Both the public and private funding sectors can help create more
of that infrastructure, she opined. These intermediaries also can leverage
public resources to get the most out of public investments.

On this topic, Linda Ortenzo, director of STEM programs at the Carnegie Science Center in Pittsburgh, described the Chevron Center for STEM
Education and Career Development, which takes into account all of the
major STEM programs in the area and includes a teacher excellence academy.[4] She said this effort has created a group of people who represent all
of those stakeholders in the STEM learning system. It also has created a
process for schools to evaluate their STEM education programs and figure
out how to increase integration, which it now is piloting with three different school districts. "The goal is to give schools rails to run on, not a
prescription, to allow for diversity but to give them guidance to get to the
place where we'd all like to see the whole ecosystem go," she explained.

Finally, Bevan urged that groups interested in influencing policy work
through state associations of school boards, given the difficulty of work-

---

[4]More information about the center is available at http://www.carnegiesciencecenter.org/
stemcenter [June 2014].

ing with all of the individual school boards. Such relationships can lead to policy guidance from state associations that comes from a source local boards can trust and use. "It really supports advocacy work at the local level," she said.

5

Breakout Sessions by Topic

On both days of the convocation, breakout groups met to consider particular issues and opportunities for collaboration among the informal, afterschool, and formal STEM education sectors. The topics they discussed were

- alignment of learning opportunities and creating cross-sector collaboration among schools, afterschool programs, and informal science education providers;
- innovations in pre-service and educator professional learning arising through cross-sector collaboration;
- assessment of student goals for STEM learning systems;
- online learning technologies in promoting collaboration and cross-sectional learning communities; and
- joint funding and policy solutions for cross-sectional collaboration and implementation.

In plenary sessions after each breakout session, reporters appointed by each group summarized the group's main conclusions. In some cases, topics were discussed in both breakout sessions, while in other cases just a single breakout group considered the topic.

ALIGNMENT OF LEARNING OPPORTUNITIES

The breakout group on alignment identified three major themes, said Rich Rosen, senior practice leader for STEM system design at the Teaching Institute for Excellence in STEM in Cleveland Heights, Ohio.

The first theme was that all parts of the STEM learning system have a common vision, which one breakout group defined as a focus on youth development. This common vision gives the sectors a shared goal on which they can base exchanges among themselves.

The second theme was the importance of infrastructure. An extensive infrastructure that can support collaboration already exists, said Rosen. To implement something new, pieces of the necessary infrastructure likely already exist, and the breakout participants noted "there's no reason to layer a new infrastructure on top." Instead, asset mapping of the existing infrastructure could identify resources that are already available.

The third theme was the need to develop a common language. In particular, if the sectors were able to agree on goals and measures, such as a badging system for STEM achievement or the development of inquiry skills, all parts of the learning system could be focused on these goals. Intermediaries can help different sectors communicate and collaborate without confusion over terms, according to the group's discussions.

Given these three themes, the group discussed two ideas to move from talking to doing, Rosen stated. The first is to build on the activities of individuals and organizations that communities already value and trust. The second is not to plan an activity all the way to the end but to get it started and then let it develop organically. "Let the process not be a one size fits all," said Rosen. "Let the process unfold, and see which things naturally bubble to the top."

PRE-SERVICE AND EDUCATOR PROFESSIONAL LEARNING

The breakout groups on pre-service and educator professional learning arising through cross-sector collaborations discussed many existing models and approaches that could be replicated and expanded, said Jim Kisiel, associate professor at California State University, Long Beach, and Joan Bissell, director of teacher education and public school programs in the California State University Chancellor's Office, who reported for the two groups. These innovations involve different actors doing different things, whether a university and local community are working on teacher professional development or an afterschool program is working with a local school. But they both said a common theme is that many of these interactions provide a way for pre-service teachers to learn more about teaching in settings in addition to schools.

For example, as Bissell noted, the American Museum of Natural

History in New York City is the only museum in the nation that has a graduate program and grants certifications. It offers a number of experiences to teachers, analogous to a medical residency model, to enable them to work with different student groups and in different settings. Other examples noted included afterschool leadership programs, an online registry of certified training programs, collaborative professional development, and pre-service teachers taking over a classroom to enable a teacher to meet with afterschool staff.

The breakout groups on this topic discussed opportunities for such interactions to increase, both in number and kind. Various obstacles can obstruct these interactions, including funding, human capacity, and time, and the groups said these obstacles need to be overcome to scale up these interactions. Bissell noted that pre-service education also has entered a time of great opportunity for innovation. For example, as a result of recent legislation in California, lifting the one-year cap on credential program length, teacher candidates can begin their training with an internship in a summer program before they learn in a more traditional way about teaching (California Senate, 2013; Sawchuck, 2013). Research has shown that teachers who start in this way tend to teach in a different way than those students who progress through more traditional pre-service programs. For example, they are more likely to use the investigative and active practices that they have seen in afterschool or summer programs in their own teaching. These programs are also valuable as a recruiting tool for students or professionals who may be considering teaching. Making these kinds of training resources online and accessible would offer a way to scale up this approach.

One prominent topic of conversation in the breakout groups was the need to identify the attributes that make a program or activity successful, both reporters observed. Though learning systems differ from place to place and time to time, identifying commonalities among these attributes could provide a list of best practices for other innovators, according to the groups.

Something else that would be extremely useful, Bissell noted, is an online glossary of terms for use in cross-sector collaborations. Such a glossary would make it possible for everyone to know what a particular term means. A second need he suggested is the ability to certify mastery of competencies, with competencies developed by experts and rigorous methods for candidates to demonstrate their attainment of those competencies.

ASSESSMENT OF STUDENT GOALS

The breakout group on assessment noted that several recent developments have changed the discussion about testing and assessment. A widespread sense exists that assessment in schools is broken, said Gil Noam, professor at Harvard University, and that something originally intended to help students learn better has instead become a source of anxiety and confusion. One question is whether afterschool assessments should differ from those given during the school day. A deeper question is whether the types of learning that students need can be rigorously assessed. "Can one connect the real learning, the real fun, the real exploration of children, and the real outcomes of their learning to a form of assessment?" Noam queried.

The situation may seem dismal, said Noam, but he said that is not accurate. In fact, many interesting and creative assessments exist that are addressing these issues. "We need to build on those rather than come out of these meetings saying we need assessments," he said. "There's a lot to build on."

The breakout group also discussed data sharing, which has been an obstacle in the past because of privacy concerns. But school districts around the country are now overcoming these problems so that the elements of the STEM learning system can share information. "Boston is one example," said Noam, "where school districts are willing to open up the information."[1]

Finally, individual students' outcomes are important in assessment, but another important factor to assess is the context, said Noam, asking "What are good programs, what are the contexts, what are the learning strategies, who are the people who are actually teaching children across contexts?" He said answering such questions would yield outcomes at different levels that are not only broad and apply across systems but also in depth on particular factors.

ONLINE COLLABORATION

The first working group on online mechanisms for collaboration settled on three main themes, reported David Greer, executive director with the Oklahoma Innovation Institute in Tulsa.

One was the need for collaborative matching when sharing online curricula, Greer explained. Besides providing a means for exchanging information, technology can help identify who is out there, who is doing what, and which efforts communicate most beneficially. Rather than dupli-

---

[1]For more information, see http://www.pearweb.org [June 2014].

cating resources, online exchanges can enable the leveraging of existing resources and the creation of collaborative resources for the future. Also, online technologies have been used in the past to replicate fairly mediocre educational techniques, Greer said. Rather than using technology to do what has always been done, the group noted that teaching could be used to present material in a way that captivates this generation of children.

The second theme was the sustainability of initiatives. "How can you not only start these programs but keep them going effectively?" he said. The group suggested that to be cost-effective, content needs to be sustainable, with modules maintained and upgraded on a regular basis. But today's students are accustomed to media-rich content with advanced simulations and graphics, which, as Greer pointed out, can be expensive.

The third theme identified by the breakout group was the need for a culture shift in thinking about how to approach the integration of STEM learning across sectors. Rather than decrying a problem or praising a solution, American educators need to ask what is working and what is not working, said Greer. "It's okay to celebrate the successes, but let's also talk about what's not working and why it's not working so we can progress," he said.

Participants in a second breakout group on online resources talked about the "Amazon model," reported David Evans, executive director of the National Science Teachers Association, in which additional resources are suggested based on what a teacher has used in the past. In this way, educators could build their own professional development portfolio, as well as enhancing interactions and sharing of resources across communities. For students and for teachers, online resources also could play a role in a badging system for student achievement and for professional development, the group suggested. Also, in their use of online resources, both teachers and students generate a great deal of data, and Evans said the possible uses of these data have not been well explored.

Scaling up either learning opportunities or professional development requires online approaches, said Evans. That can be the difference between reaching millions of people and thousands. But, he noted, resources need to be vetted and curated, rather than just consisting of the first three hits of a Google search. On the other hand, excessive control over resources can result in others setting the agenda or constraints on resources because of intellectual property protections. Collection points, portals, or hubs could make vetted materials accessible. Another possible model is what Evans called the "Yelp model," where the users of resources rate those materials and help determine whether they are recommended to others.

This breakout group also discussed blended learning, where online resources are combined with face-to-face learning. This approach, too, raises question of scale, though face-to-face interactions can be facilitated online through telepresence technologies, as the group explored.

Educators cannot be afraid to fail, Greer observed. He said, "The engineering process in general is designed to fail so you can learn from that and progress. But our education system doesn't seem to be built that way. We're afraid to fail in anything, but how are we going to learn and how are we going to make bigger steps to improve opportunities for our kids?"

Online technologies represent a tremendous resource for educators from different sectors, along with a way to communicate and exchange ideas. However, Evans also noted that teaching is a clinical process and not one that will be replaced by online resources that are doled out in a mechanical way. "These are resources to help people who interact personally with children," he said.

JOINT FUNDING AND POLICY SOLUTIONS

Kathleen Traphagen reported on the breakout session on policy and funding approaches for cross-sector collaborations. First, she said, the work of intermediary organizations is often critical to the success of cross-sector collaboration, yet it has been difficult to find funding for these organizations. It also has been difficult to fund evaluation work, especially since practitioners tend not to have enough knowledge, expertise, or time to do such evaluations effectively, she observed.

Private philanthropies provide the most flexibility and are the most willing to take risks, breakout session participants discussed. On this note, Gerald Solomon, executive director of the Samueli Foundation, called attention to the work of the STEM Funders Network, which consists of a diverse group of education-focused philanthropies. Though the funders are very different, they share a common interest in and commitment to the improvement of STEM learning. They have come together to do some things that none could do alone, including sponsoring this convocation and activities to implement the ideas developed at the convocation. "We want to be able to translate some of our experiences to be able to help communities all across the country in building the types of integrated networks that we're talking about," said Solomon.

The practitioners in the breakout group noted that it can be very problematic when funders switch their evaluation strategies in the middle of a project, Traphagen continued. Also, when grant-making priorities change, programs need enough lead time to be able to think how to replace that funding.

For both funders and projects, being honest about the risks and challenges is good policy, said Traphagen. She pointed out that good communications, trust, being upfront about expectations, and recognizing when a match is not working are also important. Jay Labov, a staff member at the National Research Council and the primary staff organizer for the convocation, mentioned that he had heard in a number of conferences that researchers had expressed the desirability of aligning reporting requirements for grants. Organizations with grants from multiple funders spend an enormous amount of time and money trying to accommodate different funders. Common metrics for programs could increase efficiency while also providing for accountability. "It's a win-win situation," he said.

At the policy level, people and organizations tend to be focused on particular parts of the STEM learning system, not on the whole system, but Traphagen suggested that a brief, coherent, and potent message could direct attention to the entire system and not just one of its parts. Perhaps the new findings of the rapidly developing learning sciences could provide such a message, said Traphagen, or the need to be a life-long STEM learner given the pace of change in the modern world. As an example, Traphagen quoted a line from a colleague, Sam Houston: STEM stands for "Strategies That Engage Minds."[2]

Policies could be shaped to support a national effort on STEM learning systems around a core set of strategies and principles, Traphagen said. These systems may look different in different communities, but they would have common objectives. For example, the California STEM Learning Network[3] has developed a policy agenda that connects the implementation of the Next Generation Science Standards with informal education and prioritizes innovative STEM networks. Not every state will be able to implement such policies, but leverage points exist in each state, Traphagen noted. Of particular importance, federal standards apply in each state, and these standards can influence state and local policies.

Finally, she said, a STEM learning system can provide educators with opportunities to be what they wanted to be when they entered education. It allows them creativity and opportunities to experience the joy of learning with children. This is a good leverage point, said Traphagen, because it builds grassroots support for participating in cross-sector collaborations. Funders could add to this support by creating multicity programs or demonstrations of effective STEM learning in and out of schools.

---

[2]This slogan was first used by the North Carolina STEM Education Center (http://ncsmt. org/ [June 2014]). For more information, also see Atkinson et al. (2013).
[3]More information is available at http://cslnet.org/ [June 2014].

6

Breakout Sessions by Sector

In the third and final breakout session of the convocation, the representatives of each of the three sectors were asked to meet separately and discuss how the theme of *STEM Learning Is Everywhere* might apply to the formal, informal, and afterschool sectors. Each breakout group was asked to address the following questions:

- How could they interact with the members of other sectors?
- What could be measurable outcomes of those interactions six weeks, six months, and one year in the future?
- How could online resources contribute to these outcomes?

In the convocation's final plenary session, reporters for the three breakout groups then recounted their groups' observations and conclusions.

THE INFORMAL SECTOR

The group from the informal sector spent much of its time developing a single transformative idea, said Margaret Honey, who reported for the group. Today, the conversation about education dwells largely on testing and evidence. But the developers of assessments have not been able to produce a high-quality standardized measure that can be used broadly across contexts. The way out of this dilemma, said Honey, is to change the conversation by producing "a very different vision of what counts as learning." This learning would be inquiry-based, student-centered, self-reflective, rigorous, and comprehensive. The evidence base would consist of portfolios documenting student learning, not a single number on a test.

Many resources would be available to achieve this objective, such as mapping of the field and new social media tools. But the more general resource is that the informal sector has the opportunity to engage young people in what Honey called "a very expansive notion of learning."

As Elizabeth Stage said, in following up on Honey's presentation, "it's the vision of science learning that needs to change. . . . If we could change that vision, we could leverage all the public investment and private and independent investment . . . to be a game changer." California is especially well positioned to lead this movement, Stage continued, because the legislature and governor would look favorably on such a plan and funding support is available. In this way, California could act as a pilot for other states in fostering cross-sector collaboration.

THE AFTERSCHOOL SECTOR

The afterschool sector has much to contribute to cross-sector collaboration, said Ellie Mitchell, director of the Maryland Out of School Time Network. "The theme of the conversation was doing what we do best so we can be successful jointly," she said.[1]

A working group of key stakeholders from each of the three sectors would provide an opportunity to learn more about each other and frame future work, said Mitchell. This working group then could be represented at both formal and informal convenings, such as the meetings of the National AfterSchool Association,[2] the National Science Teachers Association,[3] and the Midwest Afterschool Science Academy.[4] In addition, it could reach out to the National Research Council, National Science Foundation, and other organizations to discuss inclusion of cross-sector partnership language in solicitations and on review panels.

In the longer term, the group could define a message and framework for collaboration and start surfacing potential outcomes specific to the afterschool sector, Mitchell reported. Vision statements, issue briefs, and other materials could focus on the contributions of afterschool STEM programming and collaboration among sectors.

The breakout group suggested that one aspect of cross-sector collabo-

---

[1]The PowerPoint file used for this reporting session is available at http://www.samueli. org/stemconference/documents/After_School_Sector_Perspectives_on_Action_Items.pdf [June 2014].

[2]More information is available at http://naaweb.org [June 2014].

[3]More information is available at http://nsta.org [June 2014].

[4]More information is available at http://projectliftoff.net/curriculum/MASA%204.0 [June 2014].

ration could be to analyze the Common Core State Standards and Next Generation Science Standards to identify elements where the afterschool sector is best poised to collaborate and support common efforts. In addition, opportunities for joint professional development could have both immediate and long-term benefits in generating new knowledge through research and development. "Afterschool is a place for innovation," said Mitchell, "and we could be leaders in that R&D effort."

Finally, these collaborative efforts need a "clarion call" that expresses the new vision of STEM learning, Mitchell said, with champions leading the effort but everyone contributing.

THE FORMAL SECTOR

Cross-sector research around network improvement communities can help the entire STEM learning system move forward collectively, said Christina Trecha, director of the San Diego Science Project at the University of California, San Diego.[5] These networks can in turn provide the support that will enable progress to be sustainable, she said.

In addition, communications between the informal and formal sectors could increase awareness of the need for an institutionally supported culture of informal professional educators. "The formal education sector is really interested in having that conversation," said Trecha. Higher education could offer a bridge between the informal and formal sectors, both in its preparation of educators and by working with current practitioners. For example, credentialing for educators in both the formal and informal sector could help create a supportive culture of professionalism.

In the short term, the group suggested that the formal sector could reach out to current or potential partners from different sectors and begin new conversations around collaboration to meet regional needs in STEM education. The result could be leadership teams that define action steps and templates for institutional reforms.

In the medium term, the sectors could work on a strategic plan with an explicit goal of using research-based evaluations to share lessons learned locally with larger STEM networks. This could represent an initial cycle of inquiry, Trecha said. "'Trying things on' is a term that we use in the San Diego Science Project: Trying things on, seeing if they work, and trying them on again," she commented.

In the long term, cross-sector collaborations could lead to new activities or approaches that engage multiple sectors, from the scale of a single

---

[5]The PowerPoint file for this reporting out session is available at http://www.samueli. org/stemconference/documents/Formal%20Sector_Perspectives_on_Action_Items.pdf [June 2014].

teacher to a statewide network, Trecha noted. No matter what the scale of this engagement, the experiences resulting from this implementation phase would provide lessons from which all can learn, even teachers and school systems without ready access to such networks.

Finally, the breakout group suggested the establishment of an online searchable database to identify working models, similarities among approaches, and new innovations.

7

Comments from Convocation Participants

Throughout the convocation, participants had opportunities to comment on the issues raised by presenters and breakout groups. This final chapter of the summary compiles these comments as a way of revisiting and elaborating on the major themes of the convocation.

AN ECOSYSTEMS APPROACH

A prominent topic of conversation was the usefulness of the ecosystem analogy for STEM learning systems. Margaret Honey, president and chief executive officer of the New York Hall of Science, emphasized the importance of specifying what an "ecosystem" means in the context of a STEM learning system. In ecology, an ecosystem is a complex system that unfolds in interesting ways at different times, she noted. If the emphasis is on measuring what is happening here and now, important information will be lost. Assessments can measure some things at some times, but the informed observations of educators and learners can be even more valuable, she said. In particular, students can be powerful judges of their own interests, knowledge, and motivations.

Carol Tang, program officer with the S.D. Bechtel, Jr., Foundation, further developed the analogy. As she described, an ecosystem consists not just of the top-level carnivores, but also of decomposers such as bacteria and fungi, and sometimes the critical species in an ecosystem is not a sea otter but a sea star. Similarly, a largely overlooked community organizer or leader may end up being the one who sparks significant change, she suggested. Also, ecosystems are not efficient, they evolve over very long time periods, and they constantly change. "When we think about this work and try to make it too easy and clean, we are losing the messiness that is inherent in the system and that makes it beautiful and resilient," Tang said.

Several participants mentioned the role of diversity in ecosystems. Diversity is necessary for an ecosystem to thrive, said Anita Krishnamurthi, vice president of the Afterschool Alliance, and collaborative efforts need to take care not to diminish the strengths of individual sectors. She also pointed out that diversity can be intimidating, because it implies that there are many options facing programs. Bronwyn Bevan, director of the Research and Learning Institute at the Exploratorium, responded that education systems tend to simplify and move in single directions, so a case can be made for diversity of approaches. As an example of the diversity with which STEM learning systems must deal, Bernadette Chi, interim director of the Coalition for Science After School, noted that different types of organizations provide different types of programming. Some are very focused on STEM subjects, while others are not STEM focused at all, she pointed out.

Ryan Collay, director of the Science and Math Investigative Learning Experiences Program at Oregon State University, pointed to the need to identify common goals while respecting the diverse cultures of the parts of the STEM learning system. Even within individual sectors, subgroups can have varying perspectives, despite the "universal effort to do good." Ecosystems are marked by resilience, diversity, energy flow, and other attributes. "How do we see that interconnectivity? What is that wonderfully messy collective program?" he asked. Research can do much to untangle influences on the success of programs, he said, which in turn will help make the effort sustainable in the long term.

An important aspect of ecosystems, said Harry Helling, president of Crystal Cove Alliance, is that they have external drivers, which in the case of the STEM learning system include funders, policies, media communications, and community attributes. They also have internal drivers, such as strategies, assessments, and features of a system that make it more robust. An analysis that focuses on these external and internal drivers could help shape tactics and goals, he suggested.

Several participants emphasized the roles of particular players within complex and diverse ecosystems. Dennis Schatz, senior vice president at the Pacific Science Center, reminded the group about the importance of family and peers in addition to the three sectors being discussed at the convocation. He also said that he tends to think of formal education as occurring both in school and in many locations out of school. Separating afterschool or informal programs into separate sectors risks "soiling" those efforts from formal education, he said.

Bevan pointed to the importance of funders, noting that the convocation marked the first time that multiple funders with an interest in the issue had gotten involved in the conversation. "That's really important and exciting," she said.

Martin Storksdieck, a staff member at the National Research Council, emphasized the importance of including policy makers in the conversation. Private funders are important, but they cannot fund the entire agenda themselves. One way to involve policy makers is to link STEM education to other objectives, such as healthy and livable communities, he said. On that note, Honey pointed to the formation in Congress of a bipartisan STEAM caucus—for science, technology, engineering, mathematics, and design/art—that shares the same concerns as the people at the convocation.[1]

Finally, Storksdieck pointed to the analogy between learning systems and the United Nations' Agenda 21 action plan,[2] which has been ongoing for more than two decades. This movement has produced considerable experience with changing complex systems that could benefit educators and education policy makers.

THE ROLES OF EDUCATORS

The critical importance of teachers and other educators was another major topic of participants' comments.

Joan Bissell, director of teacher education and public school programs in the California State University Chancellor's Office, urged that pre-service internship placements in an informal science environment or an afterschool environment become the norm. "That's something we can do institutionally, but it also requires changes in the way certification agencies consider field experiences and clinical placement," she said. In addition, it requires that experts in informal and afterschool science join university faculties in clinical positions. "That's a major transformation," she said, "and it recognizes that expertise resides in many individuals."

Frank Pisi, director of the California AfterSchool Network, noted that the network focuses on joint professional development between school-day and expanded learning programs. The network also can influence school board members and superintendents, creating a consistent mes-

---

[1]More information is available at http://stemtosteam.org/events/congressional-steam-caucus [June 2014].

[2]More information is available at http://sustainabledevelopment.un.org/index.php?page=view&nr=23&type=400 [June 2014].

sage from both the top and the bottom. Pre-service teachers soon will be thoroughly versed in the Common Core and Next Generation Science Standards, Pisi noted, which will further support efforts to implement these new approaches in schools.

David Greer, executive director of the Oklahoma Innovation Institute, stressed the potential value of having STEM professionals work with STEM teachers. For example, STEM professionals can demonstrate to students why it is important to learn STEM subjects, beyond doing well in a test. "Maybe that's a good way to start thinking about scaling up opportunities for kids to connect with STEM," he suggested.

Kenneth Hill, president and chief executive officer of the Chicago Pre-College Science and Engineering Program, noted that the American Association for the Advancement of Science has initiated a mentorship program to work directly with teachers and schools, which is a model that could be considered for collaboration.[3]

Claudia Walker, a fifth-grade teacher from the Murphey Academy and a member of the organizing committee, emphasized the importance of reaching out not only to teachers who are enthusiastic, but also to those who are reluctant. "We need to make sure that instead of saying, 'I'm not going to do that,' they're given the opportunity to say, 'Oh, that wasn't so bad. Maybe I can,'" she said. Mike Town, science teacher at the Redmond STEM School and cochair of the organizing committee, also noted that many teachers in the formal sector are concerned about teacher evaluations based on assessments. A new vision of science education needs to overcome this concern, he said.

Hillary Salmons, executive director of the Providence Afterschool Alliance, asked how teachers can be empowered to facilitate inquiry. She expressed concern that the ways in which the Common Core standards are being implemented could displace efforts to teach students how to become self-improving problem solvers. She stated, "Let's be persistent about finding a way to enable our informal and formal educators to explore good practice around inquiry, because I think inquiry is the hardest thing in the world to do."

Jim Kisiel, associate professor of science education at California State University, Long Beach, pointed to the role of universities as the institutions that not only prepare teachers, but also educate afterschool and informal science leaders. The culture of the university therefore can have a powerful influence on collaborations and cooperation, in his view.

---

[3]More information is available at http://www.aaas.org/page/senior-scientists-and-engineers-stem-volunteer-program [June 2014].

RURAL SCHOOLS

Several participants called attention to the issues facing rural schools in engaging in cross-sector collaboration.

Margo Murphy, a high school science teacher at Camden Hills Regional High School in Maine, noted that such schools usually do not have ready access to universities or to a wealth of informal and out-of-school programs. Teachers in these settings can feel overwhelmed by new demands such as those embodied in the Next Generation Science Standards, she said, while, at the same time, support from informal and out-of-school programs, perhaps through virtual partnerships, can make such demands seem much more attainable.

Many students in rural districts who identify with STEM do so because of the efforts of a single teacher, observed Susan Kunze, a second-grade teacher at Bishop Elementary School in California. Initiatives in these areas may not provide as much "bang for the buck," said Kunze, but they are still important given the number of STEM professionals who come from rural areas.

RESOURCES

Several participants described resources that will be available to facilitate cross-sector collaboration.

Katherine Ward, an Advanced Placement biology and biotechnology teacher at Aragon High School in California, called attention to the focus on practices in the Common Core and Next Generation Science Standards. These practices explicitly state what capacities a student should be able to demonstrate, and these capacities can frame the discussion for cross-sector collaboration. "It's no longer about parsing out what I do versus what you do. It's about where do we overlap and how do we get students to this point," she said.

Phil LaFontaine, director of the Professional Learning Support Division in the California Department of Education, reminded the group of the need to keep working on exactly what STEM education means in the context of all three sectors. He also observed that parents and other members of the public need to be informed of how STEM education is changing from what they experienced so that they do not react negatively to those changes.[4]

Christopher Roe, chief executive officer of the California STEM Learn-

---

[4]Several weeks after the convocation ended, the National Academy of Engineering and the National Research Council jointly released a video titled "What Is STEM?" to address these issues. The video is available at https://www.youtube.com/watch?v=AlPJ48simtE [June 2014].

ing Network, cited the National Research Council report *Monitoring Progress Toward Successful K-12 STEM Education* (National Research Council, 2013), which developed 14 indicators around STEM education. He noted that California considered these indicators very carefully as it was choosing the metrics most relevant to that state, after which it added other metrics appropriate for the state.

Margaret Ashida, executive director of the STEMx Network, pointed to a framework for action and accountability being developed by Washington STEM that, in a single page, lays out the key elements that need to be addressed in cross-sector STEM collaboration from the early grades through the workforce, including workforce, policy, benchmarks, and high-impact strategies.[5] "Let's contribute and build on those kinds of efforts which are under way at the multistate level even as we are working for local innovation around the country," she said.

Finally, on the topic of resources, Chi emphasized the potential contributions of design thinking.[6] Aspects of the design process such as prototyping, evaluation, and "creative competence" could inform the development of STEM learning systems, she suggested. Jan Morrison, president and chief executive officer of the Teaching Institute for Excellence in STEM and Envision Excellence in STEM, echoed this idea of using design thinking. The STEM Funders Network was created using the design process, she noted, "and there's a long and good track record of using STEM to create STEM education."

Elizabeth Stage, director of the Lawrence Hall of Science, emphasized the importance of publicizing successes. She said that creating a public event that features local officials and policy makers "costs a lot of staff time but not a lot of cash" and can get "a lot of attention."

ALLIES

Finally, many convocation participants discussed the importance of developing allies. Jennifer Peck, executive director of the Partnership for Children and Youth and cochair of the organizing committee, suggested the need to have messengers outside of the STEM education field—and particularly the leaders of K-12 education—communicating with policy makers and the public about the potential of STEM learning systems. "We

---

[5]A draft of the executive summary for this framework is available at http://www.washingtonstem.org/STEM/media/Media/Our%20Approach/WA-STEM-Framework-Validation-Executive-Summary-with-Appendices.pdf [June 2014]. The final document should be available by fall 2014.

[6]As an example of this approach, see http://www.designthinkingforeducators.com [June 2014].

have a growing number of those allies, and I think this work is going to help us find more," she stated.

Cindy Hasselbring, special assistant to the Maryland state superintendent for education, emphasized the critical role of parents. Students come to kindergarten with varying degrees of preparation to learn about science, which is just one indication of how important parents are to STEM learning systems, she noted, adding that "trying to help parents understand what they can do to help support their student in STEM is a really important part of the work we are doing."

References

Afterschool Alliance. (2011a). *Afterschool: A Vital Partner in STEM Education*. Washington, DC: Author.

Afterschool Alliance. (2011b). *STEM Learning in Afterschool: An Analysis of Impact and Outcomes*. Washington, DC: Author.

Afterschool Alliance. (2013). *Defining Youth Outcomes for STEM Learning in Afterschool*. Washington, DC: Author.

Alberts, B.A. (2010). An education that inspires. *Science, 330,* 427.

Alberts, B.A. (2013). Prioritizing science education. *Science, 340,* 249.

Atkinson, J., Ralls, S., Ross, T., and Houston, S. (2013). *Strategies That Engage Minds: Empowering North Carolina's Economic Future*. Research Triangle Park: North Carolina Science, Technology, Engineering, and Mathematics Education Center.

Azevedo, F.S. (2011). Lines of practice: A practice-centered theory of interest relationships. *Cognition and Instruction, 29*(2), 147-184.

Banks, J.A., Au, K.H., Ball, A.F., Bell, P., Gordon, E.W., Gutiérrez, K.D., Heath, S.B., Lee, C.D., Lee, Y., Mahiri, J., Nasir, N.S., Valdés, G., and Zhou, M. (2007). *Learning in and out of School in Diverse Environments: Life-Long, Life-Wide, Life-Deep*. Seattle, WA: The LIFE Center (The Learning in Informal and Formal Environments Center), University of Washington, Stanford University, SRI International, and the Center for Multicultural Education.

Bransford, J.D., Vye, N.J., Stevens, R., Kuhl, P., Schwartz, D., Bell, P., Meltzoff, A., Barron, B., Pea, R., Reeves, B., Roschelle, J., and Sabelli, N. (2006). Learning theories and education: Toward a decade of synergy. In P. Alexander and P. Winne (Eds.), *Handbook of Educational Psychology* (Volume 2). Mahwah, NJ: Erlbaum.

Bronowski, J. (1956). *Science and Human Values*. New York: J. Messner.

California Senate. (2013). *Teacher Credentialing*. Bill 5. Sacramento, CA: Author. Available: http://legiscan.com/CA/text/SB5/2013 [June 2014].

Carnevale, A.P., Smith, N., and Melton, M. (2011). *STEM*. Washington, DC: Georgetown University Center on Education and the Workforce.

Change the Equation. (2013). *Lost Opportunity: Few U.S. Students Participate in STEM Out-of-School Programs*. Washington, DC: Author. Available: http://changetheequation.org/sites/default/files/CTEq%20Vital%20Signs%20Lost%20Opportunity.pdf [June 2014].

Emdin, C. (2011). Dimensions of communication in urban science education interactions and transactions. *Science Education, 95*(1), 1-20.

Kelly, D., Xie, H., Nord, C.W., Jenkins, F., Chan, J.Y., and Kastberg, D. (2013). *Performance of U.S. 15-Year-Old Students in Mathematics, Science, and Reading Literacy in an International Context: First Look at PISA 2012*. Washington, DC: National Center for Education Statistics.

Langdon, D., McKittrick, G., Beede, D., Khan, B., and Doms, M. (2011). *STEM: Good Jobs Now and for the Future*. ESA Issue Brief #03-11. Washington, DC: U.S. Department of Commerce.

National Academy of Engineering. (2008). *Grand Challenges for Engineering*. Available: http://www.engineeringchallenges.org [June 2014].

National Academy of Engineering and National Research Council. (2014). *STEM Integration in K-12 Education: Status, Prospects, and an Agenda for Research*. Committee on Integrated STEM Education, M. Honey, G. Person, and H. Schweingruber (Eds.). Washington, DC: The National Academies Press.

National Governors Association Center for Best Practices and Council of Chief State School Officers. (2010). *Common Core State Standards for Mathematics and English Language Arts*. Washington, DC: Author.

National Research Council. (2009). *Learning Science in Informal Environments: People, Places, and Pursuits*. Committee on Learning Science in Informal Environments, P. Bell, B. Lewenstein, A.W. Shouse, and M.A. Feder (Eds.). Board on Science Education, Center for Education, Division of Behavioral and Social Sciences and Education. Washington, DC: The National Academies Press.

National Research Council. (2012a). *A Framework for K-12 Science Education: Practices, Crosscutting Concepts, and Core Ideas*. Committee on Conceptual Framework for the New K-12 Science Education Standards. Board on Science Education, Division of Behavioral and Social Sciences and Education. Washington, DC: The National Academies Press.

National Research Council. (2012b). *Education for Life and Work: Developing Transferable Knowledge and Skills in the 21st Century*. Committee on Defining Deeper Learning and 21st Century Skills, J.W. Pellegrino and M.L. Hilton (Eds.). Center for Education, Board on Testing and Assessment, Division of Behavioral and Social Sciences and Education. Washington, DC: The National Academies Press.

National Research Council. (2013). *Monitoring Progress Toward Successful K-12 STEM Education: A Nation Advancing?* Committee on the Evaluation Framework for Successful K-12 STEM Education. Board on Science Education, Board on Testing and Assessment, Division of Behavioral and Social Sciences and Education. Washington, DC: The National Academies Press.

National Research Council. (2014a). *Convergence: Facilitating Transdisciplinary Integration of Life Sciences, Physical Sciences, Engineering, and Beyond*. Committee on Key Challenge Areas for Convergence and Health. Board on Life Sciences, Division on Earth and Life Studies. Washington, DC: The National Academies Press.

National Research Council. (2014b). *Developing Assessments for the Next Generation Science Standards*. Committee on Developing Assessments of Science Proficiency in K-12, J.W. Pellegrino, M.R. Wilson, K.A. Koenig, and A.S. Beatty (Eds.). Board on Testing and Assessment, Board on Science Education, Division of Behavioral and Social Sciences and Education. Washington, DC: The National Academies Press.

National Science Board. (2014). *Science and Engineering Indicators 2014*. Arlington, VA: National Science Foundation.

National Science Foundation. (2013). *Women, Minorities, and Persons with Disabilities in Science and Engineering: 2013.* Arlington, VA: Author.

NGSS Lead States. (2013). *Next Generation Science Standards: For States, By States.* Washington, DC: The National Academies Press.

Pellegrino, J.W. (2013). Proficiency in science: Assessment challenges and opportunities. *Science, 340,* 320-323.

Penuel, W.P., Lee, T., and Bevan, B. (2014). *Designing and Building Infrastructures to Support Equitable STEM Learning Across Settings.* The Research+Practice Collaboratory. San Francisco: The Exploratorium.

Sawchuck, S. (2013). California lifts one-year cap on teacher-prep programs. *Education Week.* Available: http://www.edweek.org/ew/articles/2013/09/06/03california.h33.html [June 2014].

Tai, R.H., Liu, C.Q., Maltese, A.V., and Fan, X. (2006). Planning early for careers in science. *Science, 312,* 1143-1144.

Traphagen, K., and Traill, S. (2014). *How Cross-Sector Collaborations Are Advancing STEM Learning.* Los Altos, CA: Noyce Foundation.

Varmus, H., Klausner, R., Zerhouni, E., Acharya, T., Daar, A.S., and Singer, P.A. (2003). Grand challenges in global health. *Science, 302,* 398-399.

Appendix A

Convocation Agenda

Arnold and Mabel Beckman Center
Irvine, CA
February 9, 2014-February 11, 2014

CONVOCATION GOALS AND OBJECTIVES

1. **Define the problem** . . . with more strategic, integrated approaches to STEM learning across learning platforms (informal, after-school, and formal)
2. **Identify the challenges and opportunities** . . . associated with developing a learning ecosystem
3. **What are the key attributes and characteristics** . . . for possible prototypes of strategic collaborations to move forward?
4. **Disseminate the prototypes** . . . for community uses
5. **Secure attendee commitments** . . . to work on these issues for the ensuing 18 months moving forward with plans of action

Sunday, February 9

7:00 PM	Irvine Marriott Catalina Room	Registration and Reception (with light hors d'oeuvres and beverages)

Monday, February 10

7:00 AM	Hotel Entrance	**Shuttles Depart for Beckman Center**

7:20 AM	Atrium & Dining Room	**Registration and Breakfast Buffet**

8:30 AM   Main Auditorium

Welcoming Remarks
Mike Town (Organizing Committee
    Cochair), Redmond High School
Gerald Solomon, Samueli
    Foundation, STEM Funders
    Network
Jay Labov, National Research Council

8:40 AM   Main Auditorium

Presentation: *Moving Forward with STEM Learning for All Children: What Will It Take?*
Bruce Alberts, Past President,
    National Academy of Sciences and
    Former Editor-in-Chief, *Science*

9:00 AM   Main Auditorium

Exploring the Challenges and Opportunities of Creating Local STEM Learning Collaboration Models
During this session, authors of two recently published reports on STEM learning across multiple platforms will share their findings and engage the attendees in conversation on the implications of their reports on the development of a STEM learning ecosystem.
Moderator: Martin Storksdieck,
    Board of Science Education,
    National Research Council

9-9:30 AM
STEM Learning Ecosystems
Kathleen Traphagen and Saskia
    Traill, coauthors, Noyce
    Foundation

9:30-10 AM
STEM Integration in K-12
Education: Status, Prospects, and an
Agenda for Research
Margaret Honey (NRC Committee
    Chair), New York Hall of Science

10-10:30 AM
The Myths and Challenges of
Integrating Afterschool Platforms
Anita Krishnamurthi, (Member,
    Organizing Committee),
    Afterschool Alliance
Claudia Walker (Member, Organizing
    Committee), Murphey Traditional
    Academy

Q&A with Audience

10:30 AM   Atrium   Break and Networking

11:00 AM   Main Auditorium **How the Findings and Implications**
                            of Research and Policy Can Foster
                            STEM Learning Everywhere

11-11:20 AM
Policy Issues Confronting the
Creation of a STEM Learning
Ecosystem
Jennifer Peck (Member, Organizing
    Committee), Partnership for
    Children & Youth

11:20-11:40 AM
Implications of Research Findings
on Cross-Setting Learning
Bronwyn Bevan, San Francisco
    Exploratorium
Moderator: Elizabeth Stage,
    Lawrence Hall of Science

11:40-Noon

Q&A with Audience

Topics for initial discussion:
* *Innovations in pre-service and educator professional learning arising from cross-sector collaboration*
* *Aligning learning opportunities and creating cross-sector collaboration among informal, afterschool and formal learning providers*

12:00 PM	Main Auditorium	Instructions for Afternoon Breakout Sessions and Other Logistics
12:15 PM	Atrium	Lunch and Networking
1:15 PM	Breakout Rooms *Room assignments will be determined by sign-up sheets at the registration table	**Breakout Group Discussions:** Each breakout group will explore a particular set of issues and opportunities for collaboration in greater detail:

* *Innovations in pre-service and educator professional learning arising through cross-sector collaboration*
* *Aligning learning opportunities and creating cross-sector collaboration among schools, afterschool programs, and informal science education providers*
* *Online learning technologies in promoting collaboration and cross-sectional learning communities*
* *Other topics as suggested by participants in the final morning session*

Each breakout group will appoint a

		rapporteur to discuss its findings in brief presentations during the next session
2:15 PM	Atrium	Break and Networking
2:30 PM	Auditorium	**Panel Discussion: Rapporteurs from Breakout Sessions** **Moderator:** Laura Henriques (Member, Organizing Committee), California State University, Long Beach
2:55 PM	Main Auditorium	Instructions for Afternoon Breakout Sessions and Other Logistics
3:00 PM	Breakout Rooms * To be assigned	**Breakout Group Discussions:** Each breakout group will discuss opportunities for collaboration among the three learning platforms (informal, afterschool, formal) sectors: • *Innovations in pre-service and educator professional learning arising through cross-sector collaboration* • *Online learning technologies in promoting collaboration and cross-sectional learning communities* • *Seeking joint funding and policy solutions for cross-sectional collaboration and implementation* • *Other topics that may be generated through earlier discussions (to be announced)* Each breakout group will appoint a rapporteur to discuss its findings in brief presentations on Tuesday morning

| 4:30 PM | Beckman Center Entrance | **Adjournment** Shuttles return to hotel prior to evening activity at the Discovery Science Center |

Dinner, Networking, & Discussion
Discovery Science Center, Santa Ana

6:15 PM	Hotel Entrance	Shuttles Depart for Discovery Science Center
6:30 PM	Discovery Science Center	Cocktail Reception
7:00 PM	Discovery Science Center	Dinner and Opportunity to Explore the Science Center
8:00 PM	Discovery Science Center Auditorium	Presentation by OC STEM Initiative Partners: Discovery Science Center, THINK Together, Tiger Woods Learning Center, Orange County Department of Education
8:45 PM	Science of Hockey Exhibit	Dessert
9:30 PM	Discovery Science Center Entrance	Shuttles Depart for Hotel

Tuesday, February 11

| 7:15 AM | Hotel Entrance | Shuttles Depart for Beckman Center (For those heading for airport immediately following the convocation, please bring your luggage with you) |
| 7:30 AM | Dining Room and Terrace | Breakfast Buffet and Networking |

8:30 AM	Main Auditorium	**Framing of the Day and Outcomes** Jay Labov and Gerald Solomon Brief presentations and discussion on Day 1 "aha" moments from small group breakout sessions on collaboration opportunities, and defining the attributes for prototypes
10:00 AM	Atrium	Break and Networking
10:30 AM	Breakout Rooms	**Breakout Sessions by Community:** Participants from each learning community (informal, afterschool, formal) will meet separately for this session. Based upon what was discussed yesterday and earlier in the morning, each group should address the following challenges: • Plans for interactions with other education sectors • Activities to engage other sectors • Measurable short-term outcomes (6 weeks) • Measurable mid-term outcomes (6 months) • Measurable longer-term outcomes (1 year) • Contributions to online resources for use by all sectors Each group should appoint a rapporteur to report the group's findings in the session after lunch
12:15 PM	Atrium	Lunch and Networking
1:15 PM	Auditorium	**Brief Group Reports, Final Questions and Discussion, and Next Steps** Final thoughts from organizing committee followed by open discussion

2:30 PM	Beckman Center Entrance	**Convocation Adjourns** Transportation back to hotel and John Wayne Airport (signup sheets will be available at the registration table)

Appendix B

Convocation Attendees

Nicole Akridge
Orange County STEM Initiative
Corona Del Mar, CA

Bruce Alberts
University of California
San Francisco, CA

Elizabeth Allan
University of Central Oklahoma
Edmond, OK

Maureen Allen
Orange County Science and
    Engineering Fair
Huntington Beach, CA

Julie Angle
Oklahoma State University
Stillwater, OK

Nancy Arroyo
Ysleta Independent School District
El Paso, TX

Margaret Ashida
STEMx Network, Battelle
Latham, NY

Janet Auer
Chevron Corporation
San Ramon, CA

Arthur Beauchamp
University of California
Davis, CA

Gyla Bell
Tiger Woods Foundation
Irvine, CA

James Bell
Center for Advancement of
    Informal Science Education
Washington, DC

Bronwyn Bevan
Exploratorium
San Francisco, CA

Kathy Bihr
Tiger Woods Learning Center
Washington, DC

Joan Bissell
California State University System
Long Beach, CA

Xan Black
Tulsa Community College
Tulsa, OK

Matt Blakely
Motorola Solutions Foundation
Chicago, IL

Jeff Bradbury
Cerritos College
Norwalk, CA

CJ Calderon
Orange County STEM Initiative
Corona Del Mar, CA

Russ Campbell
Burroughs Wellcome Fund
Research Triangle Park, NC

Betty Carvellas
Essex Junction, VT

Caleb Cheung
Oakland Unified School District
Oakland, CA

Bernadette Chi
Coalition for Science After School
Berkeley, CA

Connie Chow
Science Club for Girls
Cambridge, MA

Ryan Collay
Science and Math Investigative
    Learning Experiences Program
Corvallis, OR

Beth Cunningham
American Association of Physics
    Teachers
College Park, MD

Justin Duffy
World Learning, Inc.
Washington, DC

Janet English
El Toro High School
Lake Forest, CA

David Evans
National Science Teachers
    Association
Arlington, VA

Michael Feder
National Research Council
Washington, DC

Dorothy Fleisher
W.M. Keck Foundation
Los Angeles, CA

Michelle Freeman
Samueli Foundation
Corona del Mar, CA

Michael Funk
California Department of
    Education
Sacramento, CA

Ellen Gannett
National Institute on Out-of-
    School Time
Wellesley, MA

Marilyn Garza
Santa Barbara Junior High School
Santa Barbara, CA

Margaret Glass
Association of Science-Technology
    Centers
Washington, DC

Sheryl Goldstein
Harry and Jeanette Weinberg
    Foundation, Inc.
Owings Mills, MD

David Greer
Oklahoma Innovation Institute
Tulsa, OK

Susan Hackwood
California Council for Science and
    Technology
Riverside, CA

Susan Harvey
S.D. Bechtel, Jr. Foundation
San Francisco, CA

Cindy Hasselbring
Maryland State Department of
    Education
Baltimore, MD

Heidi Haugen
Florin High School
Sacramento, CA

Harry Helling
Crystal Cove Alliance
Newport Coast, CA

Laura Henriques
California State University
Long Beach, CA

Andres Henriquez
National Science Foundation
Arlington, VA

Paula Hidalgo
Game Desk
Los Angeles, CA

Kenneth Hill
Chicago Pre-College Science and
    Engineering Program, Inc.
Chicago, IL

Margaret Honey
New York Hall of Science
Corona, NY

Kenneth Huff
Williamsville Central School
    District
East Amherst, NY

Andrea Ingram
Museum of Science and Industry
Chicago, IL

Arron Jirron
S.D. Bechtel, Jr. Foundation
San Francisco, CA

Juliana Jones
Longfellow Middle School
Berkeley, CA

Ambika Kapur
Carnegie Corporation of
  New York
New York, NY

Linda Kekellis
Techbridge
Oakland, CA

Jim Kisiel
California State University
Long Beach, CA

Kristoffer Kohl
Center for Teaching Quality
Carrboro, NC

Andrew Kotko
Mather Heights Elementary
  School
Mather, CA

Anita Krishnamurthi
Afterschool Alliance
Washington, DC

Susan Kunze
Bishop Unified School District
Bishop, CA

Jay Labov
National Research Council
Washington, DC

Phil LaFontaine
California Department of
  Education
Sacramento, CA

Mylo Lam
Game Desk
Los Angeles, CA

Jason Lee
Detroit Area Pre-College Science
  and Engineering Program
Detroit, MI

Katie Levedahl
California Academy of Sciences
San Francisco, CA

Steven Long
Rogers High School
Rogers, AR

Ellen McCallie
National Science Foundation
Arlington, VA

Jeanne Miller
Lehigh Carbon Community
College
Schnecksville, PA

Mary Miller
The University of Texas
  Austin, TX

Ellie Mitchell
Maryland Out of School Time
  Network
Baltimore, MD

Jan Morrison
Teaching Institute for Excellence
  in STEM
Cleveland, OH

Margo Murphy
Camden Hills Regional High
  School
Camden, ME

Dennis Neill
Schusterman Family Foundation
Tulsa, OK

Sue Neuen
Science@OC
Santa Ana, CA

Gil Noam
Harvard University
Cambridge, MA

Christine Olmstead
Orange County Department of
    Education
Costa Mesa, CA

Steve Olson
Professional Writer
Seattle, WA

Linda Ortenzo
Carnegie Science Center
Pittsburgh, PA

Ron Ottinger
Noyce Foundation
Los Altos, CA

Jennifer Peck
Partnership for Children and
    Youth
Oakland, CA

Karen Peterson
National Girls Collaborative
    Project
Lynnwood, WA

Angela Phillips-Diaz
California Council on Science and
    Technology
Riverside, CA

Frank Pisi
California AfterSchool Network
Davis, CA

Amber Ptak
Gill Foundation
Denver, CO

Beth Rasa Edwards
University of Missouri Extension
Blue Springs, MO

Alex Reeves
Clinton Global Initiative
New York, NY

Candace Reyes-Dandrea
New York City Department
    of Youth and Community
    Development
New York, NY

Julie Renee Shannan
Girlstart
Austin, TX

Chad Ripberger
Rutgers Cooperative Extension
Trenton, NJ

Christopher Roe
California STEM Learning
    Network
San Francisco, CA

Rich Rosen
Teaching Institute for Excellence in
    STEM
Cleveland, OH

Hillary Salmons
Providence After School Alliance
Providence, RI

Dennis Schatz
Pacific Science Center
Seattle, WA

Brian Shay
Canyon Crest Academy
San Diego, CA

Freddie Shealey
Cleveland Metropolitan School
    District
Cleveland, OH

James Short
American Museum of Natural
    History
New York, NY

Maria Simani
University of California
Riverside, CA

Christopher Smith
Boston After School and Beyond
Boston, MA

Gerald Solomon
Samueli Foundation
Corona del Mar, CA

Osvaldo Soto
Math for America
San Diego, CA

Elizabeth Stage
Lawrence Hall of Science
Berkeley, CA

Maryann Stimmer
FHI360
Durham, NC

Martin Storksdieck
National Research Council
Washington, DC

Carol Tang
S.D. Bechtel, Jr. Foundation
San Francisco, CA

Sheikisha Thomas
C.E. Jordan High School
Durham, NC

Tessie Topol
Time Warner Cable
New York, NY

Jo Topps
WestEd
Los Angeles, CA

Mike Town
Redmond STEM School
Redmond, WA

Saskia Traill
The After-School Corporation
New York, NY

Kathleen Traphagen
Consultant to STEM Funders
    Network
Amherst, MA

Christina Trecha
University of California
San Diego, CA

Deborah Vandell
University of California
Irvine, CA

Claudia Walker
Murphey Traditional Academy
Greensboro, NC

Natasha Walker
Project Exploration
Chicago, IL

Katherine Ward
Aragon High School
San Mateo, CA

Steve Weaver
After School Matters
Chicago, IL

Alison White
Ohio STEM Learning Network-
    Akron Hub
Akron, OH

Janet Yamaguchi
Discovery Science Center
Santa Ana, CA

CynDee Zandes
THINK Together
Santa Ana, CA

Ashley Zauderer
Harvard University
Cambridge, MA

Debbie Zipes
Indiana Afterschool Network
Indianapolis, IN

Appendix C

Brief Biographies of Committee Members and Presenters

COMMITTEE COCHAIRS

Jennifer Peck is the executive director of the Partnership for Children and Youth in Oakland. Since joining the Partnership in 2001, she has launched initiatives to build afterschool and summer programs, meal programs, and nutrition education programs in California's lowest-income communities. She created the California Afterschool Advocacy Alliance and the State Legislative Task Force on Summer Learning, and cochairs California's Summer Matters Campaign. In 2011, she was appointed senior policy advisor and transition team director for Tom Torlakson, state superintendent of public instruction. Prior to the Partnership, she was an appointee of President Clinton at the U.S. Department of Education.

Mike Town teaches Advanced Placement environmental science and environmental engineering and sustainable design at STEM High School in Redmond, Washington. He has been recognized with numerous awards for his teaching and was also awarded a National Science Foundation (NSF) Einstein Fellowship. He has written environmental and STEM curricula, including the Cool School Challenge, which won the Environmental Protection Agency Clean Air Award. He has worked on reports and workshops from the National Academies on climate change education, successful K-12 STEM education, and integrated STEM. He is a member of the Teachers Advisory Committee for the National Academies.

COMMITTEE MEMBERS

Margaret Gaston is president of Gaston Education Policy Associates in Washington, DC. She founded the Center for the Future of Teaching and Learning, and was appointed by the governor to the California Commission for Teacher Credentialing, serving as chair of the Legislative Committee and vice chair of the Commission. At the California Department of Education, she oversaw the School Improvement Program, Community Education, and other reform efforts. Gaston has been a teacher and administrator and is the education policy advisor to the California Council on Science and Technology.

Laura Henriques is professor of science education at California State University, Long Beach (CSULB). Prior to arriving at CSULB in 1995, she taught middle and high school physics and physical science. She has been principal investigator (PI) or co-PI on a number of grants, including the NSF Noyce Scholars Programs (I & II) and a PhysTEC grant. She directed the Young Scientists Camp and a summer science camp for youth and homeless children, and spearheaded a program to help elementary teachers and laid-off teachers earn a Foundational Level General Science credential. She is currently the president of the California Science Teachers Association.

Anita Krishnamurthi is the vice president for STEM policy at the Afterschool Alliance, where she leads efforts to advance policies, research, and partnerships to provide STEM education experiences in afterschool programs. She is an astronomer by training. Over the past decade, she has been involved in science education and outreach through a range of roles at the National Academy of Sciences, NASA, and the American Astronomical Society.

Claudia Walker has been a teacher for 22 years in North Carolina and New Jersey. For the last six years, she has taught mathematics and science at Murphey Traditional Academy in Greensboro. She is also a grant coordinator and tech team leader. She received a National Board Teacher Certification in 2003 and renewed her certification in 2013. She is the Singapore Math Coach for her school and is a member of a cohort implementing Engineering Is Elementary. The recipient of a Career Award for Science and Mathematics Teachers from the Burroughs Wellcome Fund in 2010, she has been a member of the Teacher Advisory Council since 2012.

PRESENTERS AND DISCUSSANTS

Bruce Alberts has served as editor in chief of *Science* (2008-2013) and as one of President Obama's first three United States Science Envoys (2009-2011). He holds the Chancellor's Leadership Chair in biochemistry and biophysics for science and education at the University of California, San Francisco, to which he returned after serving two six-year terms as the president of the National Academy of Sciences (NAS). Alberts is noted as an original author of *The Molecular Biology of the Cell* and has earned many honors and awards, including 16 honorary degrees. He currently serves on the advisory boards of more than 25 nonprofit institutions.

Bronwyn Bevan is director of the Exploratorium Institute for Research and Learning. She serves as principal investigator on several projects, including the NSF MSP Research+Practice Collaboratory, California Tinkering Network, and the NSF-funded Relating Research to Practice Website. She sits on the California STEM Afterschool Advisory Committee and the National Research Council's Committee on STEM Learning in Out-of-School Settings, and is coeditor of the Science Learning in Everyday Life section of the journal *Science Education*.

Katherine Bihr is the vice president of programs and education for the Tiger Woods Foundation (TWF). Prior to joining TWF, she was the principal of Vista View Middle School in Fountain Valley, California. Additionally, she served on the Superintendent's Cabinet in the Ocean View School District, providing guidance in the areas of physical education and the visual/performing arts. Bihr serves on several boards and on the Department of Education Leadership Council for University of California, Irvine and California State, Fullerton. In Washington, DC, she is chairman of the board of trustees for the Cesar Chavez Charter Schools for Public Policy.

Margaret Honey joined the New York Hall of Science (NYSCI) as president and CEO in 2008. Under her leadership, NYSCI has adopted Design-Make-Play as its signature strategy to promote STEM engagement and learning. She serves as a board member of the National Research Council's Board on Science Education and the National Science Foundation's Education and Human Resources Advisory Committee.

Gerald Solomon has served as the Samueli Foundation's executive director since 2008. Prior to the Samueli Foundation, Solomon served as CEO of Public Health Foundation Enterprises for seven years. He currently serves on many boards and committees and actively supports and works with Grantmakers in Health and Grantmakers for Education on local, regional, and national issues. His career has also included 18 years as a civil trial attorney and judge pro tem.

Elizabeth Stage is the director of the Lawrence Hall of Science, University of California, Berkeley. She has taught middle school through graduate students; done research in equity; developed, evaluated, and led curriculum and professional development programs; and worked on standards and assessments in California, nationally, and internationally.

Saskia Traill is the vice president of policy and research at TASC, a New York City-based organization redesigning learning opportunities for STEM and other disciplines. She leads research and policy efforts for TASC's ExpandED Schools, a reinvention of urban public schools that includes the integration of formal and informal science learning. She has coauthored articles, policy briefs, and reports on a range of issues, including engaging children in STEM and how to fund innovative education strategies.

Kathleen Traphagen is an independent writer and strategist. Her client portfolio includes national networks of philanthropies focused on out-of-school time and STEM learning, and local networks focused on reading proficiency (Springfield, Massachusetts) and K-12 education (Boston). She served as executive director of the Boston 2:00-to-6:00 After-School Initiative and as senior policy analyst for the Mayor's Office of Intergovernmental Relations in Boston. She is an elected member of the Amherst (Massachusetts) School Committee.

Janet Yamaguchi is the vice president of education at the Discovery Science Center in California. She has more than 30 years of experience in teaching, educational program design, curriculum development, and teacher professional development. She has written labs for Holt, McDougal Textbook Publishing Company; developed and taught an Afterschool Instruction course for the University of California, Irvine, Department of Education; and is currently serving on the Next Generation Science Standards Science Expert Panel for the California Department of Education.

CynDee Zandes worked for 39 years in the Greenfield Union School District in California. She is an educator, trainer, speaker, musician, and writer, and currently is the chief program officer for THINK Together, a nonprofit providing afterschool and educational services in Southern California. She is leading the team creating a first-in-class afterschool program, focused on high-quality STEM, wellness, career to college, and project-based learning opportunities.